Metal Removal and Recovery from Mining Wastewater and E-Waste Leachate

Thesis Committee

Thesis Promotor

Prof. dr. ir. Piet N.L. Lens
Professor of Biotechnology
UNESCO-IHE Institute for Water Education
Delft, The Netherlands

Thesis Co-Promotors

Prof. Michel Madon
University of Paris-Est
Paris, france

Dr. Giovanni Esposito, PhD, MSc
Assistant Professor of Sanitary and Environmental Engineering
University of Cassino and Southern Lazio
Cassino, Italy

Dr. Hab. Eric D. van Hullebusch, PhD, MSc
Hab. Associate Professor in Biogeochemistry
University of Paris-Est
Paris, france

Mentor

Dr. Eldon R. Rene
UNESCO-IHE Institute for Water Education
Delft, The Netherlands

Reviewers

Prof. Marie-Odile Simonnot
University of Lorraine
Nancy, France

Prof. Ajit P. Annachhatre
Asian Institute of Technology
Pathumthani, Thailand

This research was conducted under the auspices of the Erasmus Mundus Joint Doctorate Environmental Technologies for Contaminated Solids, Soils and Sediments ($ETeCoS^3$) and The Netherlands Research School for the Socio-Economic and natural Sciences of the Environment (SENSE).

Joint PhD degree in Environmental Technology

Docteur de l'Université Paris-Est
Spécialité : Science Technique de l'Environnement

Dottore di Ricerca in Tecnologie Ambientali

Degree of Doctor in Environmental Technology

Thèse - Tesi di Dottorato - PhD Thesis

Suthee Janyasuthiwong

Metal Removal and Recovery from Mining Wastewater and E-Waste Leachate

defended 3 July 2015

In front of the PhD committee

Prof. dr. ir. Piet N. L. Lens	Promotor
Prof. Michel Madon	Co-Promotor
Dr. Giovanni Esposito	Co-Promotor
Dr. Hab. Eric D. van Hullebusch	Co-Promotor
Dr. Eldon R. Rene	Mentor
Prof. Ajit P. Annachhatre	Reviewer
Prof. Marie-Odile Simonnot	Reviewer

Erasmus Mundus Joint Doctorate program in Environmental Technologies for Contaminated Solids, Soils and Sediments ($ETeCoS_3$)

CRC Press/Balkema is an imprint of the Taylor & Francis Group, an informa business

Published by:
CRC Press/Balkema
PO Box 11320, 2301 EH Leiden, The Netherlands
e-mail: Pub.NL@taylorandfrancis.com
www.crcpress.com – www.taylorandfrancis.com

ISBN 978-1-138-02949-1 (Taylor & Francis Group)

Table of Contents

Acknowledgements

First of all, I would like to express my sincere gratitude toward my promotor, Prof. Piet N. L. Lens, for his kindness to give me an opportunity to do this research under his supervision. Moreover, I would like to extend my gratitude to Dr. Giovanni Esposito and Dr. Hab. Eric D. van Hullebusch, my co-promotors, for their help during my ETeCoS^3 program. I would like to give my greatest appreciation towards my mentor, Dr. Eldon R. Rene, who was always there to help me throughout my PhD phase. With his help, encouragement and support, both in academic and non-academic paths, my thesis has become a big success.

I would like to also extend my gratitude towards two of my external thesis reviewers Prof. Ajit P. Annachhatre and Prof. Marie-Odile Simonnot for showing interest in my works. With their kind suggestions and valuable feedbacks, aimed at addressing the practical implications, the thesis contents were largely improvised. Moreover, I would like to thank Prof. Vera Susanne Rotter for her kind help in processing the e-waste materials and her constructive suggestions for my research.

I would like to thank two of the master students, Sheila Monica Phiri (deceased) and Riccardo Ugas, for their contribution towards this research. Their hard work and perseverance paved the way to publish good quality manuscripts. In addition, they also helped me to understand the practical difficulties of this research, its limitations and above all taught me - how to be a good mentor. Sheila, you are my first master student; thanks for your big heartedness and we will always remember you for that in our prayers. I also had the opportunity to spend the time when you won the MSc research competition at IHE. Thanks to you too, Riccardo, even though we had limitations to finish your project, you still kept working to the best of your abilities and successfully completed your MSc thesis.

Moreover, I would like to thanks all my colleagues for their friendship and support during tough times. I would like to thank Fred Kruis, Frank Wiegman, Ferdi Battes, Lyzette Robbemont, Peter Heerings, Berend Lolkema and Don van Galen for their help and support in the laboratory. Thanks to Rohan Jain, Purvi Jain, Erika Espinosa Ortiz and Estefanía López for their wonderful friendship which made my PhD life enjoyable. I really appreciate the time we all went out together, whether it was for a summer school or a short course or to relax. I still feel like, it was just yesterday when we were together even though we walk in different paths now. I also would like to thank Silvio Matassa and Marta Cardile who made my journey (mobility) to Italy commending.

Thank to the fellowship from Erasmus Mundus Joint Doctorate program in Environmental Technology for Contaminated Solids, Soils and Sediments (ETeCoS^3) which give me this big opportunity to finish my PhD. With their supports, my overall world view has widened and I can see how people endure in this small world and I got to know people's attitudes better (positive and negative). Thus, this PhD stint served as a valuable experience in my life/career.

Lastly, I whole heartedly thank my dear family living in Thailand who constantly provided me the kind support, motivation and encouragement, during the good and bad times I had to face. Thanks to my mother and brother who let me to prosper in my educational career and cherish my own dream which they had to endure from a distance.

Suthee Janyasuthiwong (September, 2015)

Summary (English)

Metal contamination in the environment is one of the persisting global issues since it not only disturbs the environmental quality, but also the environment and human health. The major contribution to this problem arises mainly from anthropogenic activities such as industries. Metal scarcity has become more severe lately where some elements have been predicted to be eradicated from the earth crust in several decades. Recently, researchers have focused their attention to recover these metals from the waste stream and reuse it in industrial production processes.

The use of agricultural wastes as a potential low cost adsorbent for heavy metal removal from wastewater is one of the most versatile technologies. In this study among the different adsorbents tested, groundnut shell gave high removal efficiencies with fewer requirements for further post treatment for Cu, Pb and Zn removal. Furthermore, the batch experiments on the main effects of process parameters (pH, adsorbent dosage, contact time and initial metal concentration) showed a major effect on the metal uptake and removal efficiency. For material regeneration, 0.2 M HCl was the most effective desorbing solution that did not alter the efficiency, up to three adsorption and desorption cycles.

The use of sulfate reducing bacteria (SRB) in bioreactors is another technology that can be applied for the treatment of metal contaminated wastewater. The SRB reduce sulfate into sulfide which further reacts with metals to form metal sulfide precipitates. The inverse fluidized bed (IFB) bioreactor is a configuration which shows prominence in utilizing SRB technology for metal contaminated wastewater treatment. Two IFB bioreactors were operated at different pH (7.0 and 5.0). The sulfate reducing activity (SRA) at pH 7.0 was higher than at pH 5.0, illustrating that pH is the main factor that affects SRA. However, thiosulfate showed a higher efficiency than sulfate as an alternate electron acceptor. The sulfide produced using thiosulfate as the electron acceptor was 157.0 mg/L, while only 150.2 mg/L was produced using sulfate and it required an adaptation period at pH 5.0 prior to successful operation. Moreover, the IFB showed a high efficiency for Cu, Ni and Zn removal from synthetic wastewater. The removal of Cu and Zn was more than 90% at pH 7.0 and 5.0 at an initial metal concentration of 25 mg/L. On the other hand, Ni was not removed at an initial concentration of 25 mg/L, as it exerted toxic effects towards SRB.

There are various types of metal contaminated waste streams which pose as a good candidate for metal recovery include electronics waste (e-waste). This e-waste has a high potential as secondary source of metal to recover especially base metals such as Cu, Ni and Zn. Printed circuit boards (PCBs) of personal computers were evaluated as the potential secondary source of Cu, Ni and Zn using hydrometallurgical and sulfide precipitation methods. The optimal conditions for metal leaching were 0.1 M HNO_3 with a liquid to solid ratio of 20 using PCBs of 0.5 - 1.0 mm particle size at 60 °C, which resulted in 400 mg Cu/g PCBs. With sulfide precipitation at a stochiometric ratio of 1:1 ($Cu:S^{2-}$), the recovery of Cu was very effective up to 90% from the leachate which accounted to approximately 0.41 g Cu/g PCBs, while Ni and Zn recovery were, respectively, 40% (0.005 g Ni/g PCBs) and 50% (0.006 g Zn/g PCBs) from the leachate in an upflow leaching column.

Samenvattig (Dutch)

Metaalverontreiniging in het milieu is één van de aanhoudende wereldwijde problemen, omdat het niet alleen het milieu beschadigd, maar ook de menselijke omgeving en gezondheid. Dit wordt voornamelijk veroorzaakt door antropogene activiteiten, met name industriële activiteiten. Metaal schaarste is een groot probleem geworden, waarbij voor sommige elementen wordt voorspeld dat ze volledig van de aardbodem zullen verdwijnen in de komende decennia. Onderzoekers hebben zich sinds kort gefocust op het terugwinnen van metalen uit afvalstromen. Deze metalen kunnen dan weer worden gebruikt voor andere industriële processen.

Het gebruik van agrarische afvalstromen als een potentieel lage kost adsorberend materiaal om zware metalen te verwijderen uit afvalwater is één van de meest belovende technologieën. Deze studie laat zien dat pinda schilletjes, van de verschillende geteste adsorbentia, een hogere verwijdering efficiëntie heeft met minder vereisten voor de nabehandeling voor Cu, Pb en Zn verwijdering. Bovendien laten de batch experimenten met de belangrijkste proces parameters (pH, adsorbent dosering, contact tijd en begin concentratie) een groot effect zien van de metaal opname en verwijdering efficiëntie. Voor materiaal regeneratie was 0.2M HCl de meest effectieve oplossing die de efficiëntie niet veranderde gedurende drie adsorptie en desorptie cycli.

Het gebruik van sulfaat reducerende bacteriën (SRB) in bioreactoren is een techniek die kan worden gebruikt voor het regenereren van metalen uit vervuild afvalwater. De SRB reduceren sulfaat in sulfide, dat dan weer verder reageert met metalen om een metaalsulfide neerslag te voren. De inverse gefluïdiseerde bed (IFB) bioreactor is een opstelling die er veel belovend uitziet voor het gebruik van de SRB technologie voor de behandeling van met metaal verontreinigd afvalwater. Twee IFB bioreactors werden opgezet met verschillende pH's (7.0 en 5.0). De sulfaat reducerende activiteit (SRA) bij pH 7.0 was hoger dan bij pH 5.0, wat aantoont dat de pH een belangrijke factor is voor SRA. Thiosulfaat als een alternatieve elektron acceptor laat een hogere efficiëntie zien dan sulfaat. Het geproduceerde sulfide met thiosulfaat als elektron acceptor was 157.0 mg/L terwijl 150.2 mg/L sulfide werd geproduceerd uit sulfaat. Daarnaast heeft sulfaat ook nog een aanpassingsperiode nodig bij pH 5.0 om succesvol te opereren. Bovendien laat de IFB een hogere efficiëntie zien voor Cu, Ni en Zn verwijdering uit synthetisch afvalwater. De verwijderingsefficientie van Cu en Zn was meer dan 90% bij pH 7 en 5.0 bij een begin concentratie van 25 mg/L. Aan de andere kant werd Ni niet verwijderend bij een concentratie van 25 mg/L, omdat het toxische effecten had op de SRB.

Metaal verontreinigde afvalstromen zoals elektronisch afval (ook wel e-waste genaamd) is een goede kandidaat voor metaal winning. Deze e-waste heeft grote potentie als een secundaire bron voor metaal winning, vooral voor metalen zoals Cu, Ni en Zn. In dit proefschrift zijn printplaten van computers bekeken als een potentiële secundaire bron voor Cu, Ni en Zn door middel van hydrometallurgie en sulfide precipitatie. De optimale condities voor metaal uitloging was een 0.1 M HNO_3 oplossing met een vaste deeltjes ratio van 20, waarbij gebruikt wordt gemaakt van 0.5 - 1.0 mm deeltjes bij 60 °C, wat overeenkomt met 400 mg Cu/g printplaat. Met sulfide precipitatie met een stoichiometrische ratio 1:1 ($Cu:S^{2-}$) was de winning van Cu erg effectief met 90% terugwinning in de uitgeloogde vloeistof, wat neerkomt op ongeveer 0.41 g Cu/g printplaat. Dit terwijl de terugwinning voor Ni, 40% (0.005 g Ni/g printplaat) en Zn 50% (0.006 g Zn/g printplaat) bedroeg voor een proces met een opwaardse uitloog kolom.

Sommario (Italian)

La contaminazione ambientale da metalli rappresenta una problematica di grande rilievo. L'accumulo di metalli può alterare l'ecosistema e minare la salute degli esseri umani che vi interagiscono. Il contributo maggiore a questo tipo di inquinamento proviene da attività antropiche quali quelle industriali. Allo stesso tempo, la scarsità dei metalli è un fenomeno in continua accentuazione, e già nell'arco di alcune decadi si prevede l'esaurimento totale di diversi elementi. Recentemente, il mondo della ricerca sta concentrando sempre maggiori sforzi nelle direzione del recupero dei metalli da acque reflue e di processo. L'obiettivo principale è quello di riutilizzarli come materia prima nell'ambito di svariati processi industriali.

L'uso di rifiuti agricoli come potenziali adsorbenti a basso costo per i metalli pesanti contenuti nelle acque reflue rappresenta al momento una delle tecnologie più versatili. Tra i vari adsorbenti testati nell'ambito del presente studio, i gusci di arachidi hanno mostrato le efficienze di rimozione più alte, senza richiedere ulteriori post trattamenti per la rimozione di Cu, Pb e Zn. Inoltre, esperimenti batch sull'effetto di vari parametri di processo (pH, dosaggio dell'adsorbente, tempo di contatto e concentrazione di metallo iniziale) hanno mostrato un'elevata influenza di quest'ultimi nei confronti dell'efficienza finale del processo di adsorbimento e rimozione. In termini di rigenerazione del materiale adsorbente, una concentrazione di 0.2 M HCl della soluzione desorbente si è dimostrata essere la più efficace, senza alterazioni dell'efficienza del processo di rimozione dei metalli fino a tre cicli di adsorbimento-desorbimento.

L'uso di batteri solfato-riduttori in bioreattori rappresenta un'altra tecnologia applicabile al trattamento di acque reflue contaminate da metalli. I solfato-riduttori riducono il solfato in solfito, il quale a sua volta reagisce con i metalli in soluzione per formare precipitati di solfuro. Il reattore a letto fluidizzato inverso (IFBR) è la configurazione più adatta al trattamento di acque reflue contaminate da metalli mediante batteri solfato riduttori. Due reattori IFB sono stati testati a differenti valori di pH (7.0 e 5.0). L'attività di solfato-riduzione a pH 7.0 è stata più elevata che a pH 5.0, evidenziando un'elevata influenza di tale parametro sul processo. Il tiosolfato si è dimostrato essere un accettore di elettroni più efficiente rispetto al solfato. Il solfuro prodotto mediante l'utilizzo di tiosolfato come accettare di elettroni è stato di 157.0 mg/L, mentre l'utilizzo di solfato ha permesso di produrre 150.2 mg/L di solfuro solo dopo un ulteriore periodo di adattamento a pH 5.0.

Inoltre, il reattore IFB si è dimostrato altamente efficiente nella rimozione di Cu, Ni e Zn dal refluo sintetico. Efficienze di rimozione oltre il 90% sono state raggiunte per Cu e Zn a pH 7.0 e 5.0, con una concentrazione iniziale di metalli di 25 mg/L. Al contrario, una concentrazione iniziale di 25 mg/L di Ni non è stata minimamente abbattuta, mostrando effetti tossici nei confronti dei batteri solfato-riduttori.

Le acque reflue contaminate da rifiuti elettronici (e-waste) costituiscono una promettente risorsa per il recupero dei metalli. Gli e-waste hanno un elevato potenziale in termini di recupero di metalli, particolarmente di metalli di base quali Cu, Ni e Zn. Le schede a circuito stampato (PCBs) dei personal computers sono state qui studiate come potenziale risorsa di Cu, Ni e Zn mediante l'utilizzo di metodi di precipitazione idrometallurgici o con solfuro. Le condizioni ottimali per la lisciviazione sono state 0.1 M HNO_3, temperatura di 60 °C, rapporto liquido-solido di 20 e PCBs ridotte a particelle di dimensioni di 0.5 - 1.0, risultando in una concentrazione finale di 400 mg Cu/g PCBs. La precipitazione con solfuri in rapporto stechiometrico 1:1 ($Cu:S^{2-}$) ha permesso il recupero di Cu dal lisciviato con efficienze del 90%, per un totale di circa 0.41 g Cu/g PCBs, mentre Ni e Zn sono stati recuperati rispettivamente per il 40% (0.005 g Ni/g PCBs) e 50% (0.005 g Ni/g PCBs) dal lisciviato trattato in una colonna a flusso verticale. Ciò indica come il recupero del Cu sia possibile mediante una combinazione di lisciviamento acido e successiva precipitazione tramite solfuri.

Résumé (French)

La contamination de l'environnement par des métaux est l'un des problèmes mondiaux persistants car non seulement elle perturbe la qualité de l'environnement mais également la santé humaine. La principale source de ce problème est principalement due aux activités anthropiques telles que les industries. La diminution des ressources en matière première est de plus en plus mentionnée dans l'actualité et certains éléments métalliques ne pourront plus être extrait de la croûte terrestre dans des conditions économiques acceptables d'ici quelques décennies. Récemment, des chercheurs ont concentré leur attention sur la récupération de ces métaux à partir de déchets dans le but de les réutiliser dans les processus de production industrielle.

L'utilisation des déchets agricoles comme adsorbant potentiel à faible coût pour le piégeage des métaux lourds présent dans les eaux usées est une des technologies les plus polyvalentes. Dans cette étude entre les différents adsorbants testés, la coquille d'arachide montre des rendements d'épuration élevés pour le Cu, Pb et Zn. En outre, les expériences en mode batch ont permis de déterminer les principaux effets des paramètres du procédé (pH, dose d'adsorbant, temps de contact et concentration initiale de métal) qui ont montré un effet majeur sur l'absorption des métaux et sur l'efficacité du piégeage. Pour la régénération des matériaux adsorbant, l'acide chlorhydrique (0,2 M) présente les performances de désorption la plus efficace qui n'altère pas l'efficacité du piégeage ultérieur des métaux, jusqu'à trois cycles d'adsorption et de désorption.

L'utilisation de bactéries sulfato-réductrices (BSR) dans des bioréacteurs est une autre technologie qui peut être appliqué pour le traitement des eaux usées contenant des métaux. Les BSR réduisent le sulfate en sulfure, qui réagit ensuite avec des métaux pour former des précipités de sulfures métalliques. Le bioréacteur inverse à lit fluidisé (IFB) est la configuration qui présente les plus d'avantage quand la technologie de BSR pour le traitement des métalliques contaminés des eaux usées est mise en oeuvre. Deux bioréacteurs IFB ont été mis en oeuvre à différents pH (7,0 et 5,0). L'activité des bactéries sulfato réductrice à pH 7,0 était plus élevée qu'à pH de 5,0, ce qui montre que le pH est un paramètre qui affecte l'activité des BSR. Cependant, le thiosulfate a montré une efficacité supérieure à celle du sulfate en tant qu'accepteur d'électrons alternatif. Le sulfure produit en utilisant du thiosulfate comme accepteur d'électrons était 157,0 mg /L, tandis que 150,2 mg /L a été produit en utilisant le sulfate et il a fallu une période d'adaptation à pH 5,0 avant la réussite de l'opération. En outre, l'IFB a montré sa grande efficacité pour l'élimination de Cu, Ni et Zn précédemment ajoutés dans des eaux usées synthétiques. Les efficacités d'élimination de Cu et Zn étaient plus de 90% à pH 7,0 et 5,0, à une concentration initiale de métal de 25 mg/L. D'autre part, l'élimination du Ni ne était pas réussie à une concentration initiale de 25 mg/L à cause des effets toxiques de ce métal envers les BSR.

Les déchets électroniques (D3E) sont de déchets contenant des métaux qui sont de bons candidats pour la récupération des métaux. Ces D3E ont un fort potentiel en tant que source secondaire de métaux en particulier pour Cu, Ni et Zn. Les cartes de circuits imprimés d'ordinateurs personnels ont été évalués comme source secondaire potentielle de Cu, Ni et Zn en utilisant des méthodes de précipitation hydrométallurgiques a base de sulfures métalliques. Les conditions optimales pour la lixiviation des métaux à partir de cartes de circuits imprimés ont été obtenues avec 0,1 M HNO_3 avec un rapport liquide solide de 20, une granulométrie variant de 0,5 à 1,0 mm à une température de 60 ° C, ce qui a permis de lixivier 400 mg Cu/g cartes de circuits imprimés. Avec la précipitation de sulfures métalliques à un rapport stœchiométrique de 1: 1 (Cu: S^{2-}), la récupération de Cu a été très efficace jusqu'à 90% de Cu lixivié ce qui représente environ 0,41 g Cu/g de cartes de circuits imprimés, tandis que la récupération de Ni et Zn étaient, respectivement, 40% (0,005 g de Ni/g de cartes de circuits imprimés) et 50% (0,006 g Zn/g de cartes de circuits imprimés). Ceci indique que le cuivre peut être efficacement récupéré à partir de cartes de circuits imprimés en utilisant une combinaison de lixiviation acide et la précipitation via la formation de sulfures métalliques.

Chapter 1 General Introduction

Chapter 1

Problem statement

Metals are necessary for living organisms in order to maintain homeostasis, for example zinc in zinc finger protein and iron in hemoglobin. However, the same can also establish toxicity at the high dose, which is the unique character of each individual metal. There are some of these metals which are toxic by nature, for instance lead (Pb), mercury (Hg), chromium (Cr) and cadmium (Cd). Lead is one of the well-known toxic metals which is found to cause harmful effects to humans in any measurable amount (Needleman et al., 2001). Cadmium is another toxic metal which use started to get decline recently, and was put into the European Restricted Hazardous Substances lists (Morrow, 2000). Metal contamination in the natural environment is still one of the persisting problems in many areas around the world, especially from anthropogenic sources. The anthropogenic activities, especially industrial activities, are the major contribution compared to other activities. The industrial processes require very high amounts of metal as raw materials to feed the production lines which resulted in a high load of metal waste generation. The Minamata disease in Japan which became a big issue around the year 1970's was caused from the methyl mercury (CH_3Hg^+) accumulated in environmental media (water and sediment) and aquaculture people consume everyday. However, the main source of methyl mercury comes from mercury of wastewater from the industry that got into various chemical and biochemical reactions in the environment. As a consequence, there are various discharge standards for heavy metals in wastewater set up by each stakeholder.

Another emerging issue related to the metal industry is resources depletion. The constant population growth made the demand for the basic needs and other accessories growing rapidly. As a consequence of technologies advancing and increases in the population growth, the industrial sector requires large amounts of the raw material to response to the market demand which keeps increasing annually (Magdoff, 2013). It is predicted that in 2050 the global demand of the metals will increase approximately 5 times compared to nowadays. In addition, the demand for Au, Ag, Cu, Ni, Sn, Zn, Pb and Sb will be higher than the current reservoir while Fe and Pt, which were one of the optimistic predictions, will also end up in the list of previously mentioned metals (Halada et al., 2008). Figure 1.1 is one of the examples for this phenomenon. Thus, sustainability resource management starts to become widely used worldwide.

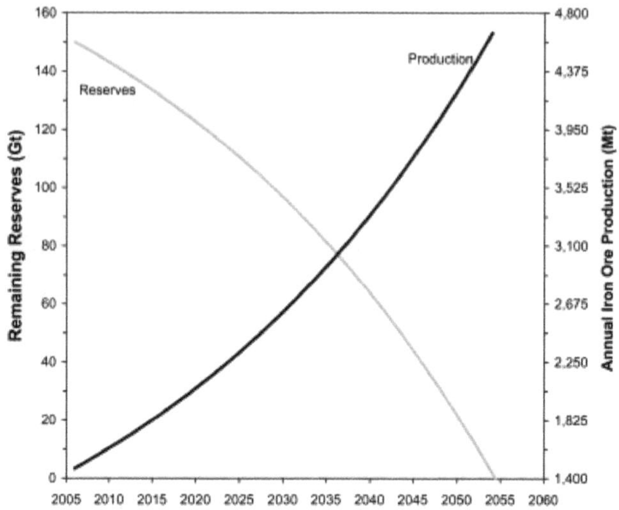

Figure 1.1 Prediction of global iron production and reservoir depletion in the period from 2005 - 2060 (Yellishetty et al., 2011)

Motivation of the study

The trends of metal consumption in the industrial sector keep increasing every year according to the demand of raw materials. The utilization rate time frame is in terms of days, while the formation rate of minerals and ore deposits can take from years to decades. Tcha and Takashina (2002) reported that the consumption pattern of seven major metals, namely steel, aluminum, copper, lead, nickel, tin and zinc is in imbalance between demand and reserve. Rauch and Pacyna (2009) also stated that the global metal cycle has changed from ores to in-use stocks. Apart from that, there is adequate evidence that shows that the high grade ore was heavily utilized and only the low grade ores are available in the reservoirs.

There are many techniques available to take the metals out from the solid phase. Hydrometallurgical methods are one of the common applications in this field. The leaching agents passes through the materials and leach the metals out in the aqueous state which will go through another process. Nowadays, many metal recovery methods from water/wastewater are available, with sulfide precipitation as one of the most famous methods. Sulfide precipitation gives advantages in term of low cost, low solubility of precipitates and ease of application. Basically, the processes involved are the addition of chemicals, mixing and settling of the precipitates. As a consequence, sulfate reducing bacteria (SRB) technologies start to gain much attention from the researchers due to their capabilities for producing sulfide under anaerobic conditions. There are many reactor configurations which have been developed for this purpose; however, it is composed of two units. The inversed fluidized bed reactor is a configuration which can reduce the need for two separated units. Villa-Gomez (2011) had showed that the IFB has a high metal removal efficiency, but the precipitates can be found as fine particles around the floating biofilm which may pose difficulties for separation. Adsorption with low cost materials, especially agricultural waste is another emerging technology which draws much attraction of researchers, especially from developing countries. Thus, the main aim of this research is to evaluate the technique for metal removal and recovery from the heavy metal contaminated waste streams.

Objectives and scope of the study

The main objective of this study is "*to investigate the feasibility of using metal contaminated wastewaters as secondary metal resource using bio-hydrometallurgical methods*".

The specific objectives are:

1. To investigate and optimize physico-chemical parameters of the adsorption efficiency of selected low cost adsorbents
2. To investigate sulfate reducing activities in IFB bioreactors under given conditions: low pH, metal contamination and thiosulfate as electron acceptor
3. To evaluate the feasibility of using IFB bioreactors and biogenic sulfide production as a mean for metal removal and recovery from wastewater
4. To study characteristic of electronic waste, specifically desktop computer PCBs, and metal leaching ability using hydrometallurgical methods
5. To evaluate the feasibility on using e-waste as secondary source of based metals, especially Cu, Ni and Zn

Outline of this thesis

This dissertation is divided into 7 chapters. The current chapter (Chapter 1) provides an overview of the study as illustrated by the flow diagram (Figure 1.2).

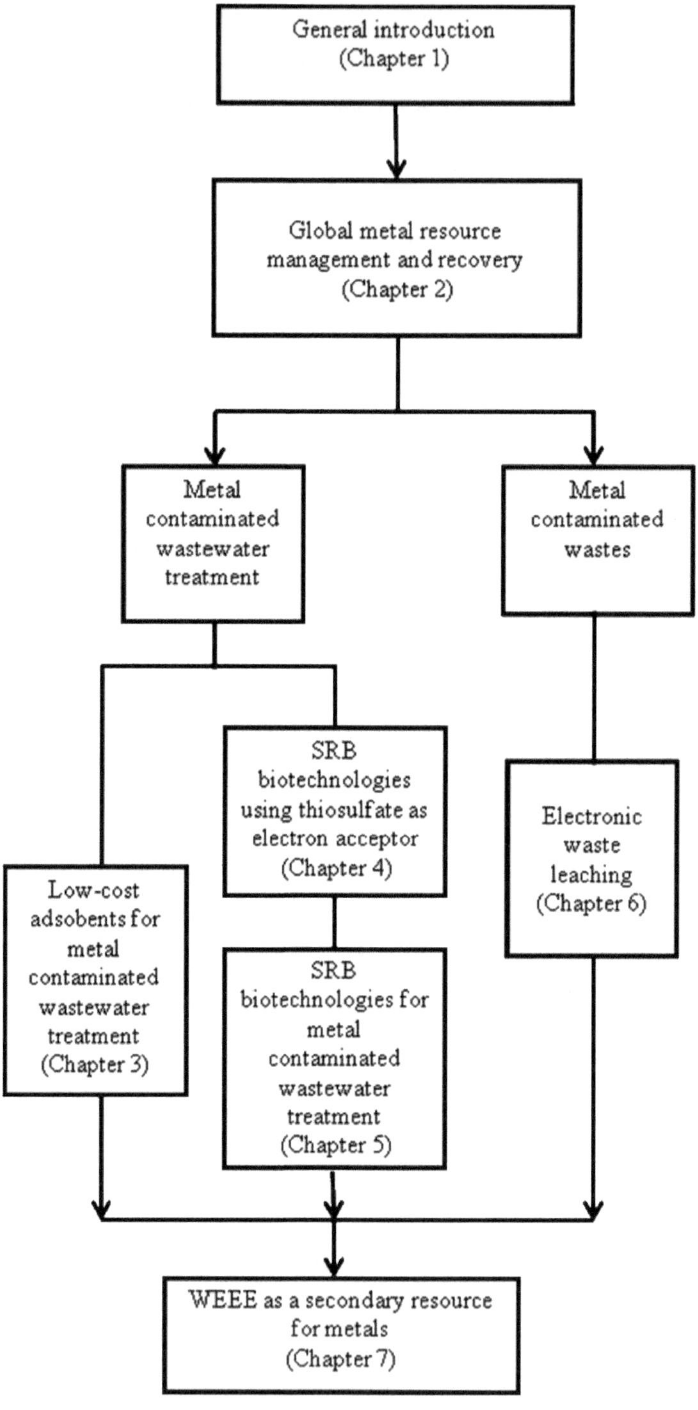

Figure 1.2 Overview of this thesis

In Chapter 2, the current global situation of metal resource management will be discussed from the cradle to grave (from mineral and ore to the end of life of metal related products). In this chapter, several examples of metal contaminated wastewater characteristics and available treatment technologies will be overviewed.

From Chapters 3 to 5, several common methods used in metal contaminated wastewater treatments will be explored. In Chapter 3, the major focus will move towards the physico-chemical methods like adsorption using low-cost adsorbents. Chapters 4 and 5 will mainly focus on the metal sulfide precipitation method using biological sulfide production from bioreactors. The SRB biotechnologies for sulfide production especially in an inverse fluidized bed (IFB) bioreactor will be demonstrated in Chapter 4. The effects of several parameters like pH and the electron acceptor on the sulfate reducing activity will be demonstrated in this chapter. Chapter 5 will display the utilization of biological sulfide production for the treatment of metal contaminated wastewater.

In Chapter 6, the major focus will shift towards the possible sources for metal recovery, especially electronic waste. The metal leachability from computer PCBs will be demonstrated in this chapter. The final chapter (Chapter 7) will elaborate the findings from the previous chapters and discusses the feasibility of using e-waste as secondary resource. The cost-benefit for large scale, environmental concerns and several issues regarding full scale applications will be discussed in this chapter.

References

Halada, K., Shimada, M. and Ijima, K., 2008. Forecasting of the consumption of metals up to 2050. Mater. Transac., 49(3), 402-410.

Magdoff, F., 2013. Global resource depletion: is population the problem? Mon. Rev., 64(8): 13-28.

Morrow, H., 2000. Cadmium and cadmium alloys, *Kirk-Othmer Encyclopedia of Chemical Technology*. John Wiley & Sons, Inc.

Needleman, H.L. and Bellinger, D., 2001. Studies of lead exposure and the developing central nervous system: a reply to Kaufman. Arch. Clin. Neuropsych., 16(4): 359-374.

Rauch, J.N. and Pacyna, J.M., 2009. Earth's global Ag, Al, Cr, Cu, Fe, Ni, Pb, and Zn cycles. Global Biogeochem. Cycles, 23(2): GB2001.

Tcha, M. and Takashina, G., 2002. Is world metal consumption in disarray? Resour Policy, 28(1-2): 61-74.

Villa-Gomez, D., Ababneh, H., Papirio, S., Rousseau, D.P.L. and Lens, P.N.L., 2011. Effect of sulfide concentration on the location of the metal precipitates in inversed fluidized bed reactors. J. Hazard. Mater., 192(1): 200-207.

Yellishetty, M., Mudd, G.M. and Ranjith, P.G., 2011. The steel industry, abiotic resource depletion and life cycle assessment: a real or perceived issue? J. Clean. Prod., 19(1): 78-90.

Chapter 2 Metal Rich Waste Stream and Feasibility of Recovery

The modified version of this chapter was submitted for publication as:

Janyasuthiwong S., Rene E.R., Esposito G. and Lens P.N.L. (2015). Metal removal and resource recovery technologies from waste stream. In: *Sustainable Technologies for Heavy Metal Removal from Soils, Solid Wastes and Wastewater*, Lens, P. N. L., Rene, E. R., Lewis, A., Sahinkaya, E. (Eds.) Book series: *Environmental Chemistry for a Sustainable World* (Eds: Lichtfouse, E., Schwarzbauer, J., Robert, D), Springer, *Under revision*

Summary

The discharge of non-treated heavy metal laden wastewater from various process industries have shown to cause detrimental effects on the environment and human health. Metal contamination in environment can be found in various forms, either as gaseous or solid or liquid phase discharge. On the other hand, metal resource depletion has also accelerated due to the advancement in technologies which consumes large amount of raw materials. Recent research topics in this field have focused on the recovery of metals from contaminated waste streams. Therefore, there are many researchers starting to focus on recovery metals from the waste streams. Currently, there are many treatment technologies for the removal and recovery of heavy metals. These technologies have their own advantages and limitations. However, ecological friendly technologies appear to have gained more attention and interest for sustainable development. The examples of these technologies are adsorption, biological sulfide precipitation and phytoremediation which utilize the inherent capability of microorganisms to treat metals under neutral conditions and produce less toxic and easily manageable end-products. These technologies have been reviewed and discussed in this chapter.

2.1 Introduction

The global demand for metals is increasing at a rapid pace due to global population increase and the expansion of industrial infrastructure in developing countries. As a consequence of this demand, metal scarcity and resource depletion have started to gain much attention in recent years. Metal is one of the common global resources which is used to produce a wide array of things/items that are aimed to comfort the daily life of humans, from construction materials or electronics to nutrient supplies which help maintain homeostasis in biological systems. The occurrence of metals in nature can be found widely from inorganic rocks and minerals to biochemical compounds like in enzymes and in protein structures. The contradiction between the utilization and the production rate had started to grow since the industrial revolution; thus, the metal reservoir became deficient. Nowadays, the fact regarding metals are very important resources which human cannot live without is growing strongly.

Due to technology advancement, the metal consumption is increasing exponentially which has directly accelerated reservoir depletion (Roberts, 1990; Yellishetty et al., 2011). The demand of metals has increased greatly in the early 2000s and the trend has shown that the growth of demand will still continue to rise. Tcha and Takashina (2002) reported that the imbalance in consumption pattern of seven major metals namely steel, aluminum, copper, lead, nickel, tin and zinc will affect the availability of these metal stocks. Similar trend in imbalance between global consumption pattern and demand was also reported by Rauch and Pacyna (2009) who stated that the global metal cycle has changed from minerals and ores to in-use stocks. In these reports, the authors have highlighted that only high grade ore was being consumed and that low grades are available in the reservoirs. Moreover, many organizations have come up with metal resource model prediction that has shown that several elements will be eliminated from the earth surface within several decades and large group will enter their endangered zone. Rosa and Rosa (2011) reported Cu productions from the Iberian Pyrite belt and found out that Cu production has passed its peak due to reservoirs depletion. Another report which supports this scenario is from Swart and Dewulf (2013). In that study, the mining data and life cycle assessment results revealed that the precious metals, especially, Au, Pd and Pt are in more critical situation compared to previous predictions that have used other aspects. Thus, there is an urgent need to ascertain solutions for metal resource management.

2.2 Metal contaminated waste stream

Although metal scarcity is a recent emerging issue, metal contamination in environment is still a persistent problem in many areas, especially, industrialized zones. Metal contamination can occur both by natural processes and anthropogenic influence. In some areas, the geography of rock and mineral contain high amount of certain elements which can be leached by surface or ground water thus resulting in high levels of a particular metal in the water bodies. The examples of this scenario are As contamination in Vietnam (Berg et al., 2001) and Bangladesh (Alam et al., 2003). These metals in the natural water can cause toxic effect to living organism if they are exposed to these pollution until the accumulated dose exceed tolerate limits. Table 2.1 shows the classification of metal toxicity based on their natural occurrence.

Although natural processes can generate metal contaminants, anthropogenic activities are the major contributions and are considered to be more powerful towards environmental degradation due to the excessive amount of HMs release within shorter time frame compared to the natural occurrences. Table 2.2 shows examples of HMs in several environmental media including living organisms which resulted from industrial activities. In this chapter, several industrial activities will be discussed.

2.2.1 Acid mine drainage

The ore excavation processes can be considered as the cradle of all anthropogenic metal activities. In order to get the raw material, the excavation of mineral and rock from the earth surface is the first step. This resource is crucial in driving nearly all human activity from daily life towards construction of convenience tools and buildings. Tilton (2013) also mentioned that the mineral wealth has high affiliation to the economics and political situation of one nation. However, the waste, especially, wastewater generated from these activities contains many hazardous pollutants such as HMs, cyanides and metalloids (Johnson and Hallberg, 2005; Sheoran and Sheoran, 2006). The waste generated from this activity can occur in various media such as soil, surface including ground water and living organisms. The acid mine drainage (AMD) is mainly generated from the oxidation of sulfide ore such as pyrite (FeS_2). The AMD is generated through both physico-chemical and biological processes (Neculita et al., 2007). The process starts from sulfide ore oxidation which gives ferrous (Fe^{2+}), sulfate (SO_4^{2-}) and proton (H^+) (equation 2.1). In some cases, the Fe^{2+} will be further oxidized under sufficient oxygen conditions and due to microbial activities to generate Ferric (Fe^{3+}), as shown in equation 2.2:

$$2\,FeS_2 + 7\,O_2 + 2\,H_2O \rightarrow 2\,Fe^{2+} + 4\,SO_4^{2-} + 4\,H^+ \quad (2.1)$$

$$4\,Fe^{2+} + O_2 + 4\,H^+ \rightarrow 4\,Fe^{3+} + 2\,H_2O \quad (2.2)$$

As shown in equation 2.3, the ferric can precipitate at pH 2.3-3.5, as hydroxide [$Fe(OH)_3$];

$$Fe^{3+} + 3\,H_2O \rightarrow Fe(OH)_3 + 3\,H^+ \quad (2.3)$$

The overall reaction from equations 2.1 to 2.3 is shown in equation 2.4 (Akcil and Koldas, 2006; Johnson and Hallberg, 2005). As the products of the reaction contain the proton, the wastewater is usually at low pH, and is very much favorable for HMs solubility.

Table 2.1 Toxicity classification of natural occurring metals (Wood, 1974)

Non toxic metals	Low toxicity metals		Moderate to high toxicity metals		
Aluminum	Barium	Praseodymium	Actinium	Lead	Tantalum
Bismuth	Cerium	Promethium	Antimony	Mercury	Thallium
Calcium	Dysprosium	Rhenium	Beryllium	Nickel	Thorium
Cesium	Erbium	Rhodium	Boron	Niobium	Titanium
Iron	Europium	Samarium	Cadmium	Osmium	Tungsten
Lithium	Gadolinium	Scandium	Chromium	Palladium	Uranium
Magnesium	Gallium	Terbium	Cobalt	Platinum	Vanadium
Manganese	Germanium	Thulium	Copper	Polonium	Zinc
Molybdenum	Gold	Tin	Hafnium	Radium	Zirconium
Potassium	Holmium	Ytterbium	Indium	Ruthenium	
Rubidium	Neodymium	Yttrium	Iridium	Silver	
Sodium					
Strontium					

Table 2.2 Metals concentration in several environmental medium as a result of anthropogenic activities

Medium	Types of waste	Metal concentration							Notes	References
		Cu	Cd	Co	Cr	Ni	Pb	Zn		
Soil (g/kg)	Abandoned Au-Ag mine	0.749	0.020				50.803	7.541		Lim et al. (2008)
	Industrial estate (tanned leather export)	0.042			2.652		0.004	0.160	Mean value	Srinivasa Gowd et al. (2010)
Sediment (g/kg)	Print factory	0.0369	0.0434				0.232	0.148	Sampling point no.6	Peng et al. (2008)
	Zn smelter	4.52	0.642				13.900	1.080	Sampling point no.8	
	WEEE processing	1.070	0.004	0.016		0.181	0.230	0.324	Lianjiang River	Wong et al. (2007b)
	Mangrove	0.056	<0.01		0.023	0.027	0.078	0.105		Defew et al. (2005)
Receiving surface water (ppb)	Print factory	14.4	4.01				4.01	73.4	Sampling point no.6	Peng et al. (2008)
	Zn smelter	20.5	74.6				10.0	941	Sampling point no.8	
	WEEE processing	7.80	0.091	0.860	2.20	36.6	1.48	30.6	Lianjiang River	Wong et al. (2007a)
Wastewater (ppm)	Metal plating	0.92			47.0	28.0		239.0	pH 2.2	Kumar et al. (2011)
		1.02			34.0	190.0		24.0	pH 2.8	
		45			44.5	394			pH 3.0	Akbal and Camcı (2011)
	Acid mine drainage	0.201				0.228	0.0155	1.28	Al: 47.9 SO_4^{2-}: 665	McCauley et al. (2009)
Landfill leachate (ppm)	filter cakes treatment of metal finishing wastes		0.17				11.20	4.40	pH 5.9	Zawierucha et al. (2013)
Organisms (see notes)	*Laguncularia racemosa* leaves	0.004					0.006	0.036	Mangrove plants (g/kg dry weight)	Defew et al. (2005)
	Common carp's gills (*Cyprinus carpio*)		1.883		0.790	1.043	1.400		in µg/g dryweight	Vinodhini and Narayanan (2008)

$$4\ FeS_2 + 15\ O_2 + 14\ H_2O \rightarrow 4\ Fe(OH)_3 + 8\ SO_4^{2-} + 16\ H^+ \quad (2.4)$$

The major reason for AMD to pose a threat to environmental quality is the treatment cost and complexity of the AMD. The low pH (~ 2.5 to 3.6) by nature is considered as an extreme condition (Nordstrom et al., 2000). Moreover, the presence of the various HMs including some metalloids such as As (Smedley and Kinniburgh, 2013) makes the decision more complicate. Therefore, universal treatment technologies are usually inadequate and too expensive in places where the mining activities occur, and therefore most of the AMD is left untreated. As a consequence, AMD has caused severe negative effects on receiving water bodies and more impact on ground water that has affected plant and animal biota (Equeenuddin et al., 2010; Yesilnacar and Kadiragagil, 2013).

2.2.2 Industrial activities

Industrial processes are the main contributors for metal contamination in environment. All production processes always generate waste. According to Table 2.2, various types of productions produce different characteristics of wastewater in terms of metal speciation and metal concentration. Thus, it is difficult to compromise with the discharge standard and treatment technologies selection. Some examples of the industries involved with the discharge of highly contaminated wastewater are fertilizer, tannery, electroplating, pulp and paper, semiconductor and electronics (Achoka, 2002; Chandra et al., 2009; Chuah et al., 2005; Soloman et al., 2009). There are also several neglected industries which can contribute to metal discharge in the environment, such as pharmaceutical and hospital (Suarez et al., 2009) because these industries are usually focused on other (hazardous) pollutants. In this section, the characteristics of several industries will be described.

2.2.2.1 Electroplating industry

Electroplating is a process in which the metal ions are moved towards the electrode for surface coating by electrical field. This process changes the subject which is being coated, in terms of its physical, chemical and also mechanical properties. Examples include, among others, corrosion resistance (chemical), appearance of coating object (physical) and hardness on the work piece surface (mechanical) (Todd et al., 1994). The process can be done by connecting the anode and cathode to the direct electrical supply with the positive terminal connected to the anode while the cathode is connected to the negative terminal. The material to be coated will be on the cathode side as shown in Figure 2.1.

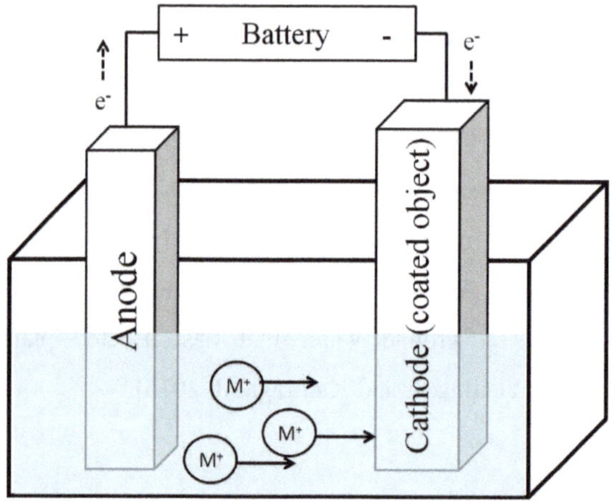

Figure 2.1 Schematic of the coating process in an electroplating industry

The metals and chemicals used in this type of industry depend on type of work piece, purpose of coating, process of coating and other factors. Even little difference in these factors may cause the chemicals and metals in the wastewater to be highly different. As a consequence, the production management and waste treatment system should be designed properly for each individual case. Several parameters of environmental importance, like pH and metal species, are shown in Table 2.3.

According to Table 2.3, the concentrations and the type of metals have wide range among each other which are highly dependent on the type of production and the processes involved in those individual plants. Majority of these plating wastewaters are quite acidic (low pH) although there are some sources which have alkaline pH. Moreover, there are several pollutants which reside along with these metals in the wastewater, such as cyanide (CN^-) and other organic solvents. The presence of these pollutants varies depending on the type of production, the quantity of usage and the processes which generate these pollutions.

Table 2.3 Characteristics of electroplating wastewater

Source	Wastewater characteristics					
	pH	Zn (mg/l)	Ni (mg/l)	Cu (mg/l)	Cr (mg/l)	Other metals (mg/l)
Chen et al. (2008)	2.0	-	94.2	-	418	-
Sirianuntapiboon et al. (2008)	12.5	-	-	27	-	-
Sousa et al. (2009)	1.3	65.5	14.4	200	-	-
Akbal and Camcı (2011)	3.0	-	394	45	44.5	-
Sankararamakrishnan et al. (2008)	11.1	-	-	-	-	Cd: 1,570.0
Kabdaşlı et al. (2009)	6.0	232	248	-	-	-
	6.0	217	270	-	-	-
	6.0	236	282	-	-	-
Kumar et al. (2011)	2.2	239	28	0.92	47	-
	2.8	24	190	1.02	34	-

2.2.2.2 Semiconductor industry

Semiconductor is a material which has electrical conductivity property, it is fundamental in modern devices nowadays. Computers, radios, televisions, solar cells, diodes and many other devices use this material. Semiconductor materials are commonly known as crystalline solids but the amorphous and liquid phase also exist. The semiconductor industry is the aggregate group of companies that designs and fabricates semiconductor devices. The growth of this industry is incredibly high due to modern technological advancements. The countries that dominate the global semiconductor market are the United States, South Korea, Japan and European Union. The revenue and the share in the market of this business are shown in Table 2.4.

According to the growth of these industries, the demand on materials also increases annually along with waste generation. The manufacturing processes of these products involve many metals and chemicals. Improper management plan on production and waste management may cause huge environmental impact. The characteristics of semiconductor wastewater from some industries are shown in Table 2.5. The major metals which can be found in this industry are Au, Cu, Pb, Ga and W.

Table 2.4 The growth of semiconductor industries in 2009 and 2010

2010 Rank	2009 Rank	Company	Country of origin	Revenue (Million USD)	Market share (%)
1	1	Intel Corporation	USA	40,020	13.2
2	2	Samsung Electronics	South Korea	28,137	9.3
3	3	Toshiba Semiconductors	Japan	13,081	4.3
4	4	Texas Instruments	USA	12,966	4.3
5	9	Renesas Electronics	Japan	11,840	3.9
6	7	Hynix	South Korea	10,577	3.5
7	5	STMicroelectronics	France/Italy	10,290	3.4
8	13	Micron Technology	USA	8,853	2.9
9	6	Qualcomm	USA	7,200	2.4
10	15	Elpida Memory	Japan	6,678	2.3

Source: iSuppli Corporation supplied rankings for 2010

Table 2.5 Characteristics of semiconductor industrial wastewater

Parameters	Source	
	Wu and Sun (2005)	Ryu et al. (2008)
pH	5.5 - 6.5	2.4 - 4.3
PSD (μm)	0.07 - 1.44	-
TSS (mg/l)	56 - 440	-
Total COD	-	221 - 444
Turbidity (NTU)	37 - 371	-
TOC (mg/l)	1.1 - 2.2	-
Conductivity (μs)	0.5 - 1.9	-
Indium (mg/l)	2.0 - 9.0	-
Phosphate (mg/l)	1.0 - 2.5	5 - 388
Total Kjeldahl nitrogen (TKN, mg/l)	-	106 -171
Chloride (mg/l)	~ 0.04	6.9 - 9.9
Calcium (mg/l)	-	6.2 - 11.6
Sulfate (mg/l)	~ 0.21	3.7 - 6.4
Sodium (mg/l)	~ 0.16	6.0 - 40.1
Potassium (mg/l)	~ 0.04	9.0 - 49.0

2.2.3 Landfill

The landfill is the place where all municipal solid waste (MSW) is gathered. It has been reported that more than 70% of the generated MSW are disposed in a landfill (Slack et al., 2007; Zacarias-Farah and Geyer-Allély, 2003) and two-thirds of this waste come from household. We can consider these places as the graves to all end-of-life products. This household waste include various types of product such as pesticide, pharmaceuticals, cleaning solution, personal care products and electronic wastes (Slack and Letcher, 2011), which can be regarded as household hazardous waste (HHW) due to its capability of releasing many toxic compounds (Slack et al., 2005). Some examples of toxic compounds which are released from the HHW are styrene, dioxins, xenobiotic organic compounds and various toxic HMs.

Countries with poor solid waste management or those countries lacking waste separation process may lead to a severe metal contamination in soil and ground water due to metal leaching. There is evidence in Guiyu, China (e-waste recycling site) where the areas and its neighborhood are heavily contaminated with the HMs. The human population living in the vicinity of this contaminated site are constantly exposed to Cd, Cr, Ni and Pb (Guo et al., 2010), while the vegetation grown in the nearby field such as rice (*Oryza sativa* L.) also shows high HMs concentration (Fu et al., 2008). One of the reasons which make landfill to become one of the possible sources for HMs contamination and transportation, especially, soil and groundwater is the pH. The landfill is mostly in the acidic condition which is generated from the degradation of organic substance, and results in the generation of volatile organic acid (VFA). Thus, this acidic condition is highly favorable for metals leaching from the solid phase. The site contamination may get more severe if there is a rainfall which facilitates the metal to reach the soil in the deeper layer or even the groundwater; for instance, landfills concentrated with HMs such as Pb (Kjeldsen et al., 2002). There are many studies that have supported the fact that the prevailing pH conditions inside the landfill has a great impact on pollution transportation, especially HMs like Cr (VI) (Haijun, 2010).

On the other hand, landfill is not only polluted with toxic metals but also metals with economical values. Morf et al. (2013) reported that the incinerator residues of MSW in Switzerland contained various metals, including precious metals (Au, Ag and Pt) and rare earth elements (REE) such as La, Ce, etc. Although the input concentration of these metals before thermal treatment in their study was still quite low, it indicates that we may lose some valuable resource during this treatment phase.

2.3 Metal removal and recovery technologies

The effluent of industries or other sources containing high level of HMs need to be treated before they are discharged into receiving surface water because HMs are toxic, carcinogenic and bio-accumulative in organisms. For instance, Pb is considered as a highly toxic metal in any measurable concentration (Needleman, 2004). There are various wastewater treatment technologies available for treating HMs contaminated wastewater prior to ultimate discharge in natural water bodies, e.g. chemical precipitation, evaporative recovery, oxidation/reduction, filtration, ion exchange, membrane technologies and electrochemical treatment technologies have been commonly used for practical applications (Fu and Wang, 2011; Liang et al., 2010). These technologies can be divided into two main categories, *viz.*, physico-chemical and biological process. Table 2.6 shows examples of metal removal efficiencies from various technologies.

2.3.1 Physico-chemical treatment

Physico-chemical technology is an effective and rapid method in dealing with metal contaminated wastewater. This treatment involves the change of physical and/or chemical properties such as size, charge and solubility. However, this method has disadvantages due to the harsh conditions used during the processes; for instance high pressure, high temperature or the addition of hazardous chemicals. Some examples of this technology are adsorption, electrocoagulation, precipitation and membrane filtration; however, in this chapter adsorption and electrocoagulation will be the main focus.

2.3.1.1 Adsorption

Adsorption has proven to be an effective physico-chemical technology for removing HMs from wastewater, especially for treating wastewaters with a low metal concentration. This technology is very effective and relatively cheap if low cost adsorbents are used. There are many adsorbents available for selection, varying from natural materials (clay balls) to waste materials (granular sludge). However, waste materials seem to attract more interest from researchers than natural materials especially agricultural waste. The advantage of using agricultural waste as HM adsorbent is the reduction of the solid waste problems and the increase in economic value and incentive of several by-products from agricultural materials. Rice husk, coconut shell, banana peel, saw dust, orange peel and groundnut shell are some examples of adsorbents from agricultural material which have been studied over the past years (Demirbas, 2008; Janyasuthiwong et al., 2015; Mohan and Singh, 2002; Orhan and Büyükgüngör, 1993; Sud et al., 2008).

Since the cost of an adsorbent depends on its abundance, availability and effectiveness, agricultural wastes have been extensively studied. In addition, these agricultural wastes are another global environment problem as burning of these wastes contributes to CO_2 emissions (Wan Ngah and Hanafiah, 2008). Thus, their use as adsorbents for wastewater treatment will reduce disposal problems (Khan et al., 2004; Lasheen et al., 2012). Besides removing metals from the wastewater, regenerating the adsorbent as well as recovering and recycling of the metals is possible using adsorption based techniques (Barrera et al., 2006; Dakiky et al., 2002). Adsorption is a method well-known for its cost effectiveness in metal removal. This method is widely used in many countries, especially developing countries and countries in transition, where they cannot afford very expensive advanced technologies. Studying the use of low cost adsorbent for HM removal is one of the major fields of research in adsorption. Researchers have focused mainly on utilizing agricultural waste materials which can be found locally or those waste materials which have no commercial values and needs to be disposed off.

2.3.1.2 Electro-coagulation

Electrocoagulation is another treatment technology applicable for the treatment of metal contaminated wastewater treatment that uses the fundamental principle of an electrochemical oxidation process. The formation of metal hydroxide flocks in the wastewater from electrodissolution is the theory behind this treatment (Akbal and Camcı, 2011). In this treatment process, the soluble anode, where the metallic cations are generated, is mainly made of aluminum or iron, while H_2 is generated at the cathode (Arroyo et al., 2009). One of the advantages of this process is that the treated water will not increase the anions or salts concentration compared to other methods, which adds chemicals for metal removal (eg: precipitation).

Tabel 1.6 Metal recovery from waste streams using various methods

Methods	Metals							Refs
	Cd	Cr	Cu	Ni	Pb	Zn	Notes	
Adsorption using natural material	-	-	-	41	-	-	Ball clays (dosage 20 g/L) pH 6.0	Chantawong et al. (2003)
	98	-	99	-	-	-	Kaolinite (dosage 20 g/L) pH 6.0	Ulmanu et al. (2003)
	-	-	66.10	-	-	45.96	Zeolites	Erdem et al. (2004)
Adsorption using biomaterial	-	-	-	96	-	-	Orang peel pH 6.0	Ajmal et al. (2000)
Adsorption using industrial waste	62.2	-	-	-	-	-	Sewage sludge pH 5.5 - 6.0	Zhai et al. (2004)
	-	-	-	78	-	-	Fry ash pH 6.5	Gupta et al. (2003)
	-	-	88.5	-	95.42	-	Iron slag pH 3.5 - 8.5	Feng et al. (2004)
Electrocoagulation	-	100%	100%	100%	-	-	Fe-Al pair electrode pH 3.0	Akbal and Camcı (2011)
	-	100%	99%	98%	-	99%	Carbon steel electrode	Al-Shannag et al. (2015)
	-	99.8%		85.6%	-	99.0%	Iron electrode pH 5.4	Kobya et al. (2014)
Precipitation	97.9%	-	98.4%	-	96%	96.5%	Sulfide precipitation SRB bioreactor	Villa-Gomez et al. (2011)
	-	-	~100	-	92	94	Sulfide precipitation	Alvarez et al. (2007)
	-	-	91.5	85.0	-	-	Sulfide precipitation	Xie et al. (2005)

Moreover, the treated effluent is clear, colorless, odorless and surprisingly low in total dissolved solids (TDS) when compared with the effluent generated from chemical treatment. In addition to that, the simplicity of the process, which requires very simple equipment and small retention times, are other incentives of using this method (Meunier et al., 2006).

Moreover, electrocoagulation is not only applicable for wastewater contaminated with metals, but includes oil and grease, suspended solids as well (Noling, 2004). This technology is quite successful in real wastewater application such as electroplating (Adhoum et al., 2004; Akbal and Camcı, 2011), tannery (Feng et al., 2007), chemical polishing (Drouiche et al., 2007; Lai and Lin, 2004) and pulp and paper wastewater (Khansorthong and Hunsom, 2009; Sridhar et al., 2011). Therefore, this technology has been used for industrial wastewater in many places such as the United States, South America and Europe (Rodriguez et al., 2007).

In terms of metals removal, there are many studies that have focused on understanding the effects of operational parameters and optimization of the processes. Examples of process parameters are pH, material and distance between the electrodes, initial metal concentration, coagulation time and current density (Bhagawan et al., 2014; Senturk, 2013; Shafaei et al., 2011). Heidmann and Calmano (2008) reported that higher initial concentrations of Cr (VI) showed higher removal rate for Cr (VI), while the removal rate of Au, Cu, Ni and Zn was not affected by this parameter. Lai and Lin (2003) demonstrated that the Al and Fe electrode pair yielded the highest performance among all of the selected pairs of electrodes during semiconductor fabrication wastewater treatment. In that study, the Cu removal was 99%, while the turbidity reduction was 96.5%. The COD of effluent was less than 100 mg/L, and the overall effluent quality was considered as good quality for reuse.

2.3.2 Biological treatment

According to Table 2.6, biological method shows nearly the same level of removal efficiency as that of physico-chemical methods but they are usually operated at milder conditions. The major disadvantages of these technologies are the time, suitability of species, the fluctuation in the performance and environmental conditions, since the process is mainly based on an organism's metabolic activity. The fundamental theory governing the biological treatment of HMs concerns the generation of alkalinity and immobilization of the HMs *via* both physical and chemical mechanisms by the organisms (Johnson and Hallberg, 2005). This technology can be divided into two main categories, *viz.*, phytoremediation and microbiological treatment. Table 2.7 compared the pros and cons of each method.

Table 2.7 Comparison of phytoremediation and microbiological treatment

Criteria	Phytoremediation	Microbiological treatment
Land availability requirement	Large area required	Not required large space
Load of pollutant	Medium - high	Medium - High
Maintenance by manpower	Hard	Easy
Manpower needed	High	Low
Effectiveness	Medium - high	High
Ecological view	Low impact	Varied
Material recovery	Need specialize	Easy

2.3.2.1 Phytoremediation

Phytoremediation is one of the biological technologies used for the treatment of pollutants present in wastewater, including HMs. This technology not only offers advantages during wastewater treatment, but also provides other advantages in terms of ecology, green area, reduce C foot print and aesthetics. The phytoremediation is the method in which selected plant species is used to mitigate the environmental problems or pollutants (metals, pesticides, solvents, crude oils and its derivatives) from soil, air, or water. There are many plant species which are commonly used in this field; *Vetiveria sp.*, *Typha sp.* and *Cyperus sp.* are examples of those plants. Maine et al. (2006) reported that the constructed wetland which were planted with several plant species; for example *Pistia stratiotes, Cyperus alternifolius and Typha domingensis*, showed the high percentage of Cr and Ni removal and the Zn concentration was below 50 μg/l. However, there was a report by Park et al. (2008) that the two wetlands, *Acorus* pond and *Typha* pond, had not shown enough efficiency on metals and metalloids removal and suggested that the fluctuations of iron and manganese concentrations were primarily due to the microbial activities under anoxic condition.

2.3.2.2 Microbiological processes (sulfate reduction)

Most microbiological processes generate alkalinity from the reduction of electron donor (C source); for example, sulfate reduction, denitrification, methanogenesis, iron and manganese reduction. However in this chapter, the focus will be mainly on sulfate reduction (SR).

Sulfate reducing bacteria (SRB) are group of bacteria which have ability to reduce sulfate to sulfide under anaerobic condition. In anaerobic respiration, sulfate is acts as an electron acceptor instead of oxygen in aerobic respiration which is the important mechanism to generate bioenergy or adenosine triphosphate (ATP). Moreover, SRB have been found to play a major role in S cycle (Figure 2.2) especially in sulfate reduction mechanism where SO_4^{2-} is reduced to S^{2-}. The most common genus mentioned in many studies was *Desulfovibrio* which are gram negative, rod shape, flagellate and strictly anaerobes. The classification of the SRB is based on several factors such as organic substrate which they utilize or incompletely degradc (Rzeczycka and Blaszczyk, 2005). For example, *Desulfovibrio desulfuricans* uses carbon dioxide (CO_2) as its carbon source and hydrogen (H_2) as its electron donor. This type of SRB which use H_2 is called autotrophic. However, in cases where CO_2 is insufficient to complete their metabolism, other organic sources of two carbon molecule compounds are required (Widdel, 1988).

Therefore, wastewater treatment technology using SRB is one of the most famous methods for the treatment of wastewater contaminated with HMs (Bijmans et al., 2011, Janyasuthiwong et al., 2015). The sulfide precipitation which is the main mechanism of this process is a very fast and effective reaction. Moreover, it also gives advantages in terms of selective metal precipitation since each metal sulfide (MS) is pH specific and the MS can tolerate a wide range of pH than hydroxide and carbonate precipitation (Lewis, 2010). Another reason which makes this technology to gain more attention is the low cost of the chemical used as sulfide source (mainly Na_2S). SRB does not only generate sulfide which is the main component for metal removal, but they also utilize the organic compound present in wastewater.

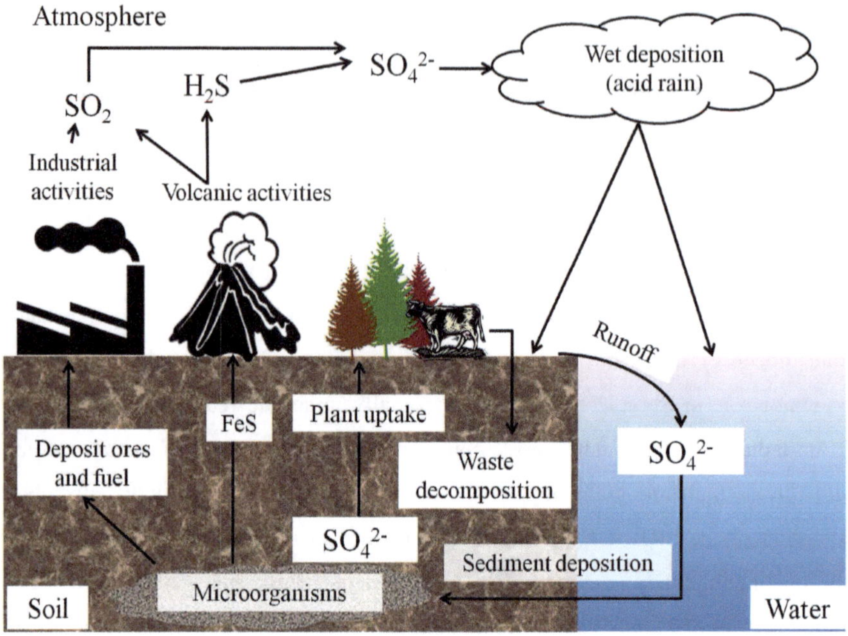

Figure 2.2 The cycle of sulfur

There are many factors involved with SR activity but pH is one of the key factors which not only limit the SR activity but also sulfide speciation (H_2S, HS^- and S^{2-}). Costa et al. (2008) reported that the growth of SRB is observed at around pH 5.0 to 7.0 and no SR activity was observed at pH 2.0. Like other living organisms, bacterial growth is also affected by many process conditions, either directly or indirectly. SRB also require specific condition in order to grow and reduce sulfate to sulfide. SRB grows at an optimum pH ranging between 5.0 and 8.0; however, there are some strains that can tolerate harsher conditions like acidophiles which can survive in AMD at pH as low as 2.0 (Neculita et al., 2007).

2.4 Conclusions

The global metal consumption rate is increasing rapidly in accordance with the exponential growth of population and advancement of production technologies. Metal contamination in the environment due to anthropogenic sources is a persistent global issue, having environmental, political and medical implications. HMs are toxic and carcinogenic and have shown to cause serious health effects on human population and the fauna and flora. As a consequence, various treatment methods have been developed for the treatment of metal contaminated waste streams. Among the commonly used physico-chemical and biological technologies for HM treatment, cost-effectiveness, technical feasibility, plant simplicity and longevity of the process are the factors that govern the selection of an appropriate technology.

References

Adhoum, N., Monser, L., Bellakhal, N. and Belgaied, J.-E., 2004. Treatment of electroplating wastewater containing Cu^{2+}, Zn^{2+} and Cr (VI) by electrocoagulation. J. Hazard. Mater., 112(3): 207-213.

Ajmal, M., Rao, R.A.K., Ahmad, R. and Ahmad, J., 2000. Adsorption studies on *Citrus reticulata* (fruit peel of orange): removal and recovery of Ni (II) from electroplating wastewater. J. Hazard. Mater., 79(1-2): 117-131.

Akbal, F. and Camcı, S., 2011. Copper, chromium and nickel removal from metal plating wastewater by electrocoagulation. Desalt., 269(1-3): 214-222.

Akcil, A. and Koldas, S., 2006. Acid mine drainage (AMD): causes, treatment and case studies. J. Clean. Prod., 14(12-13): 1139-1145.

Al-Shannag, M., Al-Qodah, Z., Bani-Melhem, K., Qtaishat, M.R. and Alkasrawi, M., 2015. Heavy metal ions removal from metal plating wastewater using electrocoagulation: Kinetic study and process performance. Chem. Eng. J., 260(0): 749-756.

Alam, M.G.M., Snow, E.T. and Tanaka, A., 2003. Arsenic and heavy metal contamination of vegetables grown in Samta village, Bangladesh. Sci. Total Environ., 308(1-3): 83-96.

Alvarez, M.T., Crespo, C. and Mattiasson, B., 2007. Precipitation of Zn (II), Cu (II) and Pb (II) at bench-scale using biogenic hydrogen sulfide from the utilization of volatile fatty acids. Chemosphere, 66(9): 1677-1683.

Arroyo, M.G., Pérez-Herranz, V., Montañés, M.T., García-Antón, J. and Guiñón, J.L., 2009. Effect of pH and chloride concentration on the removal of hexavalent chromium in a batch electrocoagulation reactor. J. Hazard. Mater., 169(1-3): 1127-1133.

Barrera, H., Ureña-Núñez, F., Bilyeu, B. and Barrera-Díaz, C., 2006. Removal of chromium and toxic ions present in mine drainage by Ectodermis of *Opuntia*. J. Hazard. Mater., 136(3): 846-853.

Berg, M., Tran, H.C., Nguyen, T.C., Pham, H.V., Schertenleib, R. and Giger, W., 2001. Arsenic contamination of groundwater and drinking water in Vietnam: A human health threat. Environ. Sci. Technol., 35(13): 2621-2626.

Bhagawan, D., Poodari, S., Pothuraju, T., Srinivasulu, D., Shankaraiah, G., Yamuna Rani, M., Himabindu, V. and Vidyavathi, S., 2014. Effect of operational parameters on heavy metal removal by electrocoagulation. Environ Sci Pollut Res, 21(24): 14166-14173.

Bijmans, M.F.M., Buisman, C.J.N., Meulepas, R.J.W. and Lens, P.N.L., 2011. Sulfate reduction for inorganic waste and process water treatment. Compr. Biotechnol.: 435-446.

Chantawong, V., Harvey, N.W. and Bashkin, V.N., 2003. Comparison of heavy metal adsorptions by Thai kaolin and ballclay. Water Air Soil Pollut., 148(1-4): 111-125.

Costa, M., Martins, M., Jesus, C. and Duarte, J., 2008. Treatment of acid mine drainage by sulphate-reducing bacteria using low cost matrices. Water Air Soil Pollut., 189(1): 149-162.

Dakiky, M., Khamis, M., Manassra, A. and Mer'eb, M., 2002. Selective adsorption of chromium (VI) in industrial wastewater using low-cost abundantly available adsorbents. Adv. Environ. Res., 6(4): 533-540.

Defew, L.H., Mair, J.M. and Guzman, H.M., 2005. An assessment of metal contamination in mangrove sediments and leaves from Punta Mala Bay, Pacific Panama. Mar. Pollut. Bull., 50(5): 547-552.

Demirbas, A., 2008. Heavy metal adsorption onto agro-based waste materials: a review. J. Hazard. Mater., 157(2): 220-229.

Drouiche, N., Ghaffour, N., Lounici, H. and Mameri, M., 2007. Electrocoagulation of chemical mechanical polishing wastewater. Desalt., 214(1-3): 31-37.

Equeenuddin, S.M., Tripathy, S., Sahoo, P.K. and Panigrahi, M.K., 2010. Hydrogeochemical characteristics of acid mine drainage and water pollution at Makum Coalfield, India. J. Geochem. Explor., 105(3): 75-82.

Erdem, E., Karapinar, N. and Donat, R., 2004. The removal of heavy metal cations by natural zeolites. J. Colloid Interf. Sci., 280(2): 309-314.

Feng, D., van Deventer, J.S.J. and Aldrich, C., 2004. Removal of pollutants from acid mine wastewater using metallurgical by-product slags. Sep. Purif. Technol., 40(1): 61-67.

Feng, J.-W., Sun, Y.-B., Zheng, Z., Zhang, J.-B., Li, S. and Tian, Y.-C., 2007. Treatment of tannery wastewater by electrocoagulation. J. Environ. Sci., 19(12): 1409-1415.

Fu, J., Zhou, Q., Liu, J., Liu, W., Wang, T., Zhang, Q. and Jiang, G., 2008. High levels of heavy metals in rice (*Oryza sativa* L.) from a typical E-waste recycling area in southeast China and its potential risk to human health. Chemosphere, 71(7): 1269-1275.

Guo, Y., Huo, X., Li, Y., Wu, K., Liu, J., Huang, J., Zheng, G., Xiao, Q., Yang, H., Wang, Y., Chen, A. and Xu, X., 2010. Monitoring of lead, cadmium, chromium and nickel in placenta from an e-waste recycling town in China. Sci. Total Environ., 408(16): 3113-3117.

Heidmann, I. and Calmano, W., 2008. Removal of Zn (II), Cu (II), Ni (II), Ag (I) and Cr (VI) present in aqueous solutions by aluminium electrocoagulation. J. Hazard. Mater., 152(3): 934-941.

Johnson, D.B. and Hallberg, K.B., 2005. Acid mine drainage remediation options: a review. Sci. Total Environ., 338(1-2): 3-14.

Khan, N.A., Ibrahim, S. and Subramaniam, P., 2004. Elimination of heavy metals from wastewater using agricultural wastes as adsorbents. Malays. J. Sci., 23(43-51): 43.

Khansorthong, S. and Hunsom, M., 2009. Remediation of wastewater from pulp and paper mill industry by the electrochemical technique. Chem. Eng. J., 151(1-3): 228-234.

Kobya, M., Erdem, N. and Demirbas, E., 2014. Treatment of Cr, Ni and Zn from galvanic rinsing wastewater by electrocoagulation process using iron electrodes. Desalin. Water. Treat.: 1-11.

Kumar, R., Bhatia, D., Singh, R., Rani, S. and Bishnoi, N.R., 2011. Sorption of heavy metals from electroplating effluent using immobilized biomass *Trichoderma viride* in a continuous packed-bed column. Int. Biodeterior. Biodegradation, 65(8): 1133-1139.

Lai, C.L. and Lin, S.H., 2003. Electrocoagulation of chemical mechanical polishing (CMP) wastewater from semiconductor fabrication. Chem. Eng. J., 95(1-3): 205-211.

Lai, C.L. and Lin, S.H., 2004. Treatment of chemical mechanical polishing wastewater by electrocoagulation: system performances and sludge settling characteristics. Chemosphere, 54(3): 235-242.

Lasheen, M.R., Ammar, N.S. and Ibrahim, H.S., 2012. Adsorption/desorption of Cd (II), Cu (II) and Pb (II) using chemically modified orange peel: equilibrium and kinetic studies. Solid State Sci., 14(2): 202-210.

Lewis, A.E., 2010. Review of metal sulphide precipitation. Hydrometallurgy, 104(2): 222-234.

Lim, H.-S., Lee, J.-S., Chon, H.-T. and Sager, M., 2008. Heavy metal contamination and health risk assessment in the vicinity of the abandoned Songcheon Au-Ag mine in Korea. J. Geochem. Explor., 96(2-3): 223-230.

McCauley, C.A., O'Sullivan, A.D., Milke, M.W., Weber, P.A. and Trumm, D.A., 2009. Sulfate and metal removal in bioreactors treating acid mine drainage dominated with iron and aluminum. Water Res., 43(4): 961-970.

Meunier, N., Drogui, P., Montané, C., Hausler, R., Mercier, G. and Blais, J.-F., 2006. Comparison between electrocoagulation and chemical precipitation for metals removal from acidic soil leachate. J. Hazard. Mater., 137(1): 581-590.

Mohan, D. and Singh, K.P., 2002. Single-and multi-component adsorption of cadmium and zinc using activated carbon derived from bagasse - an agricultural waste. Water Res., 36(9): 2304-2318.

Morf, L.S., Gloor, R., Haag, O., Haupt, M., Skutan, S., Lorenzo, F.D. and Böni, D., 2013. Precious metals and rare earth elements in municipal solid waste - sources and fate in a Swiss incineration plant. Waste Manage., 33(3): 634-644.

Neculita, C.-M., Zagury, G.J. and Bussière, B., 2007. Passive treatment of acid mine drainage in bioreactors using sulfate-reducing bacteria. J. Environ. Qual., 36(1): 1-16.

Needleman, H., 2004. Lead poisoning. Annu. Rev. Med., 55: 209-222.

Noling, C., 2004. New electrocoagulation system addresses challenges of industrial storm, wash water. Ind. Water World, 5(4): 22.

Nordstrom, D.K., Alpers, C.N., Ptacek, C.J. and Blowes, D.W., 2000. Negative pH and extremely acidic mine waters from Iron Mountain, California. Environ. Sci. Technol., 34(2): 254-258.

Orhan, Y. and Büyükgüngör, H., 1993. The removal of heavy metals by using agricultural wastes. Water Sci. Technol., 28(2): 247-255.

Peng, K., Luo, C., Lou, L., Li, X. and Shen, Z., 2008. Bioaccumulation of heavy metals by the aquatic plants *Potamogeton pectinatus* L. and *Potamogeton malaianus* Miq. and their potential use for contamination indicators and in wastewater treatment. Sci. Total Environ., 392(1): 22-29.

Rauch, J.N. and Pacyna, J.M., 2009. Earth's global Ag, Al, Cr, Cu, Fe, Ni, Pb, and Zn cycles. Global Biogeochem. Cycles, 23(2): GB2001.

Rodriguez, J., Stopić, S., Krause, G. and Friedrich, B., 2007. Feasibility assessment of electrocoagulation towards a new sustainable wastewater treatment. Environ. Sci. Pollut. Res. Int., 14(7): 477-482.

Rosa, D.R.N. and Rosa, R.N., 2011. Copper depletion in the Iberian Pyrite Belt: another indicator of global scarcity. Appl. Earth Sci., 120(1): 39-43.

Senturk, E., 2013. The treatment of zinc-cyanide electroplating rinse water using an electrocoagulation process. Water Sci. Technol., 68(10): 2220-2227.

Shafaei, A., Pajootan, E., Nikazar, M. and Arami, M., 2011. Removal of Co (II) from aqueous solution by electrocoagulation process using aluminum electrodes. Desalt., 279(1-3): 121-126.

Sheoran, A.S. and Sheoran, V., 2006. Heavy metal removal mechanism of acid mine drainage in wetlands: a critical review. Miner. Eng., 19(2): 105-116.

Slack, R. and Letcher, T., 2011. Chemicals in waste: household hazardous waste, *Waste: a handbook of waste management and recycling.* . Elsevier, Amsterdam, pp. 181-195.

Slack, R.J., Gronow, J.R., Hall, D.H. and Voulvoulis, N., 2007. Household hazardous waste disposal to landfill: using LandSim to model leachate migration. Environ. Pollut., 146(2): 501-509.

Slack, R.J., Gronow, J.R. and Voulvoulis, N., 2005. Household hazardous waste in municipal landfills: contaminants in leachate. Sci. Total Environ., 337(1-3): 119-137.

Smedley, P.L. and Kinniburgh, D., 2013. Arsenic in groundwater and the environment. In: O. Selinus (Ed.), *Essentials of Medical Geology*. Springer Netherlands, pp. 279-310.

Sridhar, R., Sivakumar, V., Immanuel, V.P. and Maran, J.P., 2011. Treatment of pulp and paper industry bleaching effluent by electrocoagulant process. J. Hazard. Mater., 186(2-3): 1495-1502.

Srinivasa Gowd, S., Reddy, M.R. and Govil, P.K., 2010. Assessment of heavy metal contamination in soils at Jajmau (Kanpur) and Unnao industrial areas of the Ganga Plain, Uttar Pradesh, India. J. Hazard. Mater., 174(1-3): 113-121.

Sud, D., Mahajan, G. and Kaur, M., 2008. Agricultural waste material as potential adsorbent for sequestering heavy metal ions from aqueous solutions: a review. Bioresour. Technol., 99(14): 6017-6027.

Swart, P. and Dewulf, J., 2013. Quantifying the impacts of primary metal resource use in life cycle assessment based on recent mining data. Resour. Conserv. Recy., 73(0): 180-187.

Tcha, M. and Takashina, G., 2002. Is world metal consumption in disarray? Resour Policy, 28(1-2): 61-74.

Tilton, J.E., 2013. *Mineral Wealth and Economic Development*. Routledge, the United States.

Ulmanu, M., Marañón, E., Fernández, Y., Castrillón, L., Anger, I. and Dumitriu, D., 2003. Removal of copper and cadmium ions from diluted aqueous solutions by low cost and waste material adsorbents. Water Air Soil Pollut., 142(1-4): 357-373.

Villa-Gomez, D., Ababneh, H., Papirio, S., Rousseau, D.P.L. and Lens, P.N.L., 2011. Effect of sulfide concentration on the location of the metal precipitates in inversed fluidized bed reactors. J. Hazard. Mater., 192(1): 200-207.

Vinodhini, R. and Narayanan, M., 2008. Bioaccumulation of heavy metals in organs of fresh water fish *Cyprinus carpio* (Common carp). Int. J. Environ. Sci. Technol., 5(2): 179-182.

Wan Ngah, W.S. and Hanafiah, M.A.K.M., 2008. Removal of heavy metal ions from wastewater by chemically modified plant wastes as adsorbents: a review. Bioresour. Technol., 99(10): 3935-3948.

Wong, C.S.C., Duzgoren-Aydin, N.S., Aydin, A. and Wong, M.H., 2007a. Evidence of excessive releases of metals from primitive e-waste processing in Guiyu, China. Environ. Pollut., 148(1): 62-72.

Wong, C.S.C., Wu, S.C., Duzgoren-Aydin, N.S., Aydin, A. and Wong, M.H., 2007b. Trace metal contamination of sediments in an e-waste processing village in China. Environ. Pollut., 145(2): 434-442.

Wood, J.M., 1974. Biological cycles for toxic elements in the environment. Science, 183(4129): 1049-1052.

Xie, Y., Xu, Y., Yan, L. and Yang, R., 2005. Recovery of nickel, copper and cobalt from low-grade Ni-Cu sulfide tailings. Hydrometallurgy, 80(1-2): 54-58.

Yesilnacar, M.I. and Kadiragagil, Z., 2013. Effects of acid mine drainage on groundwater quality: a case study from an open-pit copper mine in eastern Turkey. Bull. Eng. Geol. Environ., 72(3-4): 485-493.

Zacarias-Farah, A. and Geyer-Allély, E., 2003. Household consumption patterns in OECD countries: trends and figures. J. Clean. Prod., 11(8): 819-827.

Zawierucha, I., Kozlowski, C. and Malina, G., 2013. Removal of toxic metal ions from landfill leachate by complementary sorption and transport across polymer inclusion membranes. Waste Manage., 33(10): 2129-2136.

Zhai, Y., Wei, X., Zeng, G., Zhang, D. and Chu, K., 2004. Study of adsorbent derived from sewage sludge for the removal of Cd^{2+}, Ni^{2+} in aqueous solutions. Sep. Purif. Technol., 38(2): 191-196.

Chapter 3 Adsorption Technology for Metal Contaminated Wastewater Treatment

The modified version of this chapter was published as:
Janyasuthiwong S., Phiri S.M., Kijjanapanich P., Rene E.R., Esposito G. and Lens P.N.L. (2015). Copper, lead and zinc removal from metal contaminated wastewater by adsorption onto agricultural wastes. *Environ Technol.*: 1 - 13
DOI: 10.1080/09593330.2015.1053537

Summary

The use of agricultural wastes (groundnut shell, orange peel and rice husk) as potential cost-effective adsorbent for heavy metal removal from wastewater was evaluated. The effect of pH (2.0 - 6.0), adsorbent dosage (0.6 - 2.2 g), contact time (10 - 130 min) and initial concentration (Pb: 5 - 105 mg/L, Cu and Zn: 2.5 - 52.7 mg/L) on metal removal efficiency and uptake capacity were investigated using response surface methodology to optimize the process conditions. Groundnut shell showed a high potential to remove Cu, Pb and Zn from synthetic wastewater. . The optimum conditions obtained were: 2.5 g adsorbent with 40.7 mg/L Cu at pH 4.4 and 64 min, 2.5 g adsorbent with 196.1 mg/L Pb at pH 5.6 and 60 min contact time, and 3.1 g adsorbent with 70.2 mg/L Zn at pH 4.3 and 50 min, for Cu, Pb and Zn, respectively. The regeneration of the groundnut shell was possible for a maximum of 3 cycles using 0.2 M HCl as the desorbing solution without any significant change in the adsorbing efficiency.

3.1 Introduction

The presence of heavy metals (HMs) in water is one of the main global concerns due to their detrimental effects on the environment and public health (Fu and Wang, 2011; Kiran and Thanasekaran, 2011; Wan Ngah and Hanafiah, 2008; Wang and Peng, 2010). Rapid industrial development has increased the production and use of HMs, resulting in high concentration of HMs discharged into water bodies. Industries such as mining (Alcolea et al., 2012; Davies et al., 2011), tanneries, metal plating, fertilizer industries, battery and pesticide production, ore refineries and the paper industry (Chuah et al., 2005) are the major contributing sources of heavy metal. HMs, such as zinc (Zn), lead (Pb), copper (Cu), iron (Fe), nickel (Ni), and cadmium (Cd) are toxic, carcinogenic, persistent in nature and tend to bioaccumulate (Wan Ngah and Hanafiah, 2008), which has led to the formulation of strict regulations (such as WHO, USEPA or EU) concerning the quality of wastewaters discharged from the industries. For example, Cu and Pb concentrations should be less than 2.0 and 0.01 mg/L in mining and electroplating wastewater, as suggested by WHO, while the US EPA has set a limit of 1.3 and 0.015 mg/L, respectively (Fu and Wang, 2011; Hashim et al., 2011; Sud et al., 2008).

Several technologies have been developed for treating heavy metal contaminated wastewater prior to discharge in natural water bodies, e.g. chemical precipitation, evaporative recovery, oxidation/reduction, filtration, ion exchange, membrane technologies and electrochemical treatment technologies have been commonly used (Fu and Wang, 2011; Liang et al., 2010). Although high effectiveness can be obtained from these physico-chemical techniques, the operational and maintenance costs are high in these processes. Adsorption has proven to be an effective physico-chemical technology for removing HMs from wastewater, especially for treating wastewaters with a low metal concentration. This technology is cheap and environmentally friendly if low cost adsorbents are used. In the search for low cost adsorption materials over the past years, many researchers shifted their interests into the use of agricultural wastes, such as rice husk, coconut shell, banana peel, saw dust, orange peel and groundnut shell. Since the cost of an adsorbent depends on its abundance, availability and effectiveness, agricultural wastes have been extensively studied. In addition, these agricultural wastes are another global environment problem as burning of these wastes contributes to CO_2 emissions (Wan Ngah and Hanafiah, 2008). Thus, their use as adsorbents for wastewater treatment will reduce disposal problems (Khan et al., 2004; Lasheen et al., 2012). Besides removing metals from the wastewater, regenerating the adsorbent as well as recovering and recycling of the metals is possible using adsorption based techniques (Barrera et al., 2006; Dakiky et al., 2002).

Response surface methodology (RSM) is one of the statistical tools which can be used to model and analyze the influence of individual and interaction effects among variables in order to estimate the optimum response of statistical valuables (Landaburu-Aguirre et al., 2010). In general, this optimization is carried out by varying one factor at a time, by keeping the other variables constant. The traditional models are often time consuming and yield a low efficiency for the optimization process This approach has been tested for HMs removal with various types of materials; for instance, *Aspergillus niger* (Amini et al., 2008), microwaved olive stone (Alslaibi et al., 2013) and illite (Turan et al., 2011). The advantage of using statistically design method consists of estimating the coefficient of a mathematical model, prediction of the response and checking the accuracy of the model (Kalantari et al., 2014). There are many parameters that affect the efficiency of an adsorption process such as pH, temperature, adsorbent dosage and initial metal concentration. For instance, pH may affect the metal availability, functional groups on the surface of the adsorbents and ionic strength (Zhang, 2011). In case of adsorbent dosage and metal concentration, the efficiency mainly depends on the availability of active sites and the competition for active sites (Amuda et al., 2007). The major investigation of the present research was to screen and optimize the different low-cost absorbent materials from agricultural wastes for the removal of Cu, Pb and Zn from metal contaminated wastewater and evaluate the effects of pH, contact time, adsorbent dosage and initial metals concentration on metal uptake and removal efficiencies. The reuse ability of the selected adsorbent was also evaluated to envisage the maximum usability of the material.

3.2 Material and Methods

3.2.1 Materials

3.2.1.1 Reagents

The metals of interest in this study, viz., Cu, Pb and Zn were prepared from metal chloride salts: $CuCl_2 \cdot 2H_2O$, $PbCl_2$ and $ZnCl_2$, respectively. The desorbing solutions used during the regeneration experiments were NaOH, HCl, sodium acetate ($C_2H_3NaO_2.3H_2O$), sodium EDTA (Na_2EDTA) and $Ca(NO_3)_2$. All chemicals used were of analytical grade.

3.2.1.2 Adsorbents

The six adsorbents used in this study were rice husks (from Tanzania), coconut husk and Wawa tree saw dust (from Ghana), while banana peel, orange peel and groundnut shell were locally produced in The Netherlands. These low-cost adsorbents were prepared as followed: they were washed three times with demineralized water, dried for 24 h at 105.0°C and grounded with the blenders, and sieved to the particle size of 0.5 - 0.8 mm as shown in Figure 3.1.

3.2.2 Adsorption experiments

3.2.2.1 Material screening

All adsorbents were screened by a series of batch experiments in order to investigate the efficiency of each material on the metal removal efficiency. The factors that nominated of particular selection of material were metal removal efficiency, metal uptake capacity and dissolved organic carbon (DOC) released from the material. Experiments were carried out in 250 mL polypropylene bottles, at room temperature (25.0 ± 2.0°C). Screening was done in 2 steps; during the first step, 6 materials were screened at initial metal concentration of 10 mg/L at pH 7.0 at 100 rpm shaking speed using 1.0 g of adsorbent for a duration of 24 h. The sampling intervals were at 0, 5, 10, 20, 30, 60, 120, 180, 360, 720 and 1440 minutes, respectively. During the second screening step, the 3 selected materials (orange peel, groundnut shell and rice husk) were done in similar pattern as the first screening, but at pH of 3.0 and 5.0, respectively while the contact time was increased up to 2880 min.

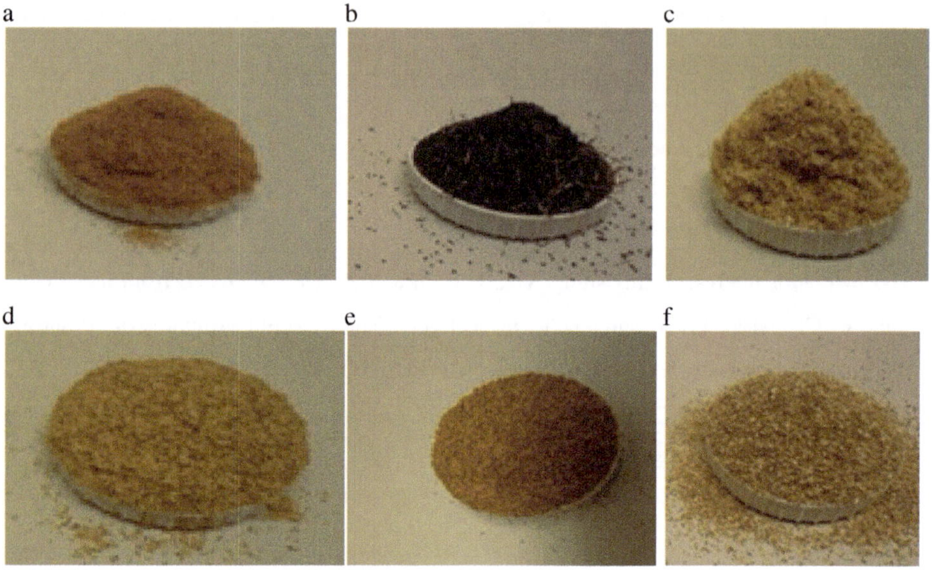

Figure 3.1 Processed adsorbents used in this study a) Coconut husk, b) Banana peel, c) Saw dust, d) Rice husk, e) Groundnut shell, and f) Orange peel

3.2.2.2 Experimental design

Batch optimization experiments of groundnut shell, the selected adsorbent, were carried out according to the statistically significant central composite design (CCD) technique. According to the factorial design (Equation 3.1), the values for each parameter were coded as - α, -1, 0, +1, and +α, respectively, where *k* is number of factors, - α is the lowest value, 0 is the average of the lowest and highest values and +α is the highest value (Table 3.1).

$$\alpha = 2^{k/4} \qquad (3.1)$$

Equation 2 represents the relationship between the actual and coded values;

$$\text{Coded values} = \frac{\left(X_i - \overline{X_i}\right)}{\Delta X} \qquad (3.2)$$

where X_i is actual value of the ith factor in the uncoded units, $\bar{X}$ the average of high and low values in the ith factor and ΔX is the step change, respectively.

Table 3.1 The range of values of the parameters used in the study

Parameters	Uncoded value				
	-α	-1	0	+1	+α
Initial Cu and Zn concentration (mg/L)	2.5	15.0	27.5	40.0	52.5
Initial Pb concentration (mg/L)	5.0	30.0	55.0	80.0	105.0
Adsorbent dosage (g)	0.6	1.0	1.4	1.8	2.2
pH	2.0	3.0	4.0	5.0	6.0
Contact time (min)	10	40	70	100	130

3.2.3 Analysis of main parameters and interaction effects

The results for metal removal efficiency, metal uptake, removal rate and DOC were calculated using Equations 3.3, 3.4, 3.5and 3.6, respectively.

$$\text{Removal efficiency } (\%) = \frac{(C_0 - C_t)}{C_0} \times 100 \qquad (3.3)$$

$$\text{Metal uptake (mg/g), } q_t = \frac{(C_0 - C_t)}{m} \times V \qquad (3.4)$$

$$\text{Removal rate } \left(\frac{mg}{g \cdot \min}\right) = \frac{(C_0 - C_t)}{m \times t} \times V \qquad (3.5)$$

$$\text{DOC (mg/L)} = \frac{V}{m} \times C \qquad (3.6)$$

Where, C_o (mg/L) is the initial metal ion concentration, C_t (mg/L) is the metal ion concentration at time t, m (g) is the dry weight of adsorbent, V (L) is the volume of aqueous solution, t (min) is contact time, and q_t (mg/g) is the adsorption capacity at time t. For Equation 6, V is the volume (mL) of water used, m is the dry weight (g) of the adsorbent, and C (mg/L) is the concentration of non purgeable organic carbon (NPOC).

3.2.3.1 Response surface methodology

A stepwise RSM procedure was applied in order to determine the effects of pH, initial metal concentrations, contact time and adsorbent dosage (Table 3.1) on the HMs removal efficiencies and uptake capacity of the adsorbent. The results obtained from the statistically designed batch experiments were analyzed using the software MINITAB 15. Gauss Jordan optimization method was used to estimate the optimum conditions from the respective regression equations (Anupam et al., 2011). ANOVA was used to estimate the linear, square and interaction effects of the parameters (pH, contact time, adsorbent dosage and initial metal concentrations) on the metal removal efficiency and the metal adsorption capacity of the adsorbent.

3.2.3.2 Optimization of the adsorbent

A model equation was developed for each of the selected metal (Cu, Pb and Zn) based on the regression coefficients obtained from statistical analysis. Equation 3.7 represents the typical second order polynomial equation derived to predict the optimum conditions for metal uptake capacity and metal removal efficiency.

$$Y_i=b_0+b_1X_1+b_2X_2+b_3X_3+b_4X_4+b_{11}X^2{}_{11}+b_{22}X^2{}_{22}+b_{33}X^2{}_{33}+b_{44}X^2{}_{44}+b_{12}X_1X_2+b_{13}X_1X_3+b_{14}X_1X_4 +b_{23}X_2X_3+b_{24}X_2X_4+b_{34}X_3X_4 \quad (3.7)$$

Where, Y_i is the predicted response, b_0 is the coefficient of the constant, b_1, b_2, b_3 and b_4 are linear coefficients, b_{11}, b_{22}, b_{33} and b_{44} are quadratic coefficients, b_{12}, b_{13}, b_{14}, b_{23}, b_{24}, and b_{34} interaction coefficients, while X_1, X_2, X_3 and X_4 are the coded independent variables, respectively (Anupam et al., 2011; Bhatia et al., 2011). The equations were solved and the values for X_1, X_2, X_3 and X_4 were obtained. The actual values for the parameters were calculated according to Equation 3.2. After optimization, batch experiments were conducted in triplicates at the optimum conditions obtained from the model. The results were then compared with the model predicted metal uptake capacity values in order to ascertain the predictive capability of the models.

3.2.4 Adsorbent regeneration experiments

The regeneration of the materials is another factor that should be considered, since it determines the life cycle and the efficiency of the material before it is needed to be replaced, in order to achieve the desire efficiency. Regeneration of groundnut shell, the selected adsorbent, was carried out by performing adsorption/desorption cycles in batch experiments. Initially, the five potential desorption solutions; NaOH, HCl, sodium acetate, sodium EDTA and calcium nitrate were screened. Then the selected desorption solution (HCl) was used for further adsorption/desorption experiments with the previously obtained optimum conditions based on metal uptake capacity. Screening experiments were carried out in duplicates, while regeneration experiments were done in triplicates. Table 2 shows the conditions for screening and regeneration experiments. At the end of each experiment, the adsorbents were carefully washed three times with demineralized water and filtered with GF/C, 110 mm Whatman filters. After ensuring that all the metal in solution phase or desorption solution has been washed out, the adsorbent was then used in the next cycle of adsorption/desorption.

Table 3.2 Experimental conditions for screening of desorption solutions and adsorbent regeneration experiments

Parameters	Experimental conditions				
	Screening experiment		Regeneration experiments		
	First screening	Second screening	Cu	Pb	Zn
Shaking speed (rpm)	100	100	100	100	100
pH	7.0	3.0, 5.0	4.4	5.6	4.3
Metal concentration (mg/L)	10	10	40.7	196.1	70.2
Contact time (min)	1440	2880	64	60	50
Desorption solution concentration (M)	-	-	0.2	0.2	0.2
Adsorbent dose (g)	1	1	2.5	2.5	3.1

3.2.5 Analytical methods

pH was measured using a 691 Metrohm pH meter and a SenTix 21 WTW pH electrode. Cu, Pb and Zn concentration were measured using an AAnalyst 200 Perkin Elmer Atomic Absorption Spectrometer (AAS) Flame (Eaton et al., 2005). DOC was measured using the high temperature combustion method by a Shimadzu TOC-V CPN analyzer (Eaton et al., 2005). All samples were filtered using 0.45 μm, 25 mm Whatman filter paper to remove all adsorbent residues.

3.3 Results

3.3.1 Screening of adsorbents

The highest metal removal efficiency was achieved in case of orange peel and groundnut shell, followed by coconut husks, rice husks and saw dust, respectively (Figure 3.2). The highest removal efficiencies obtained were 85, 99 and 92%, respectively for Cu, Pb and Zn by orange peel while 71, 99 and 68% removal efficiencies were obtained from groundnut shell. Banana peel as the adsorbent showed the lowest metal removal efficiency. However, in the case of coconut husks, a significant color change in the medium was observed and the metal-laden solutions were forming precipitates during the experiments. Hence, coconut husk was ruled out from further experimentation

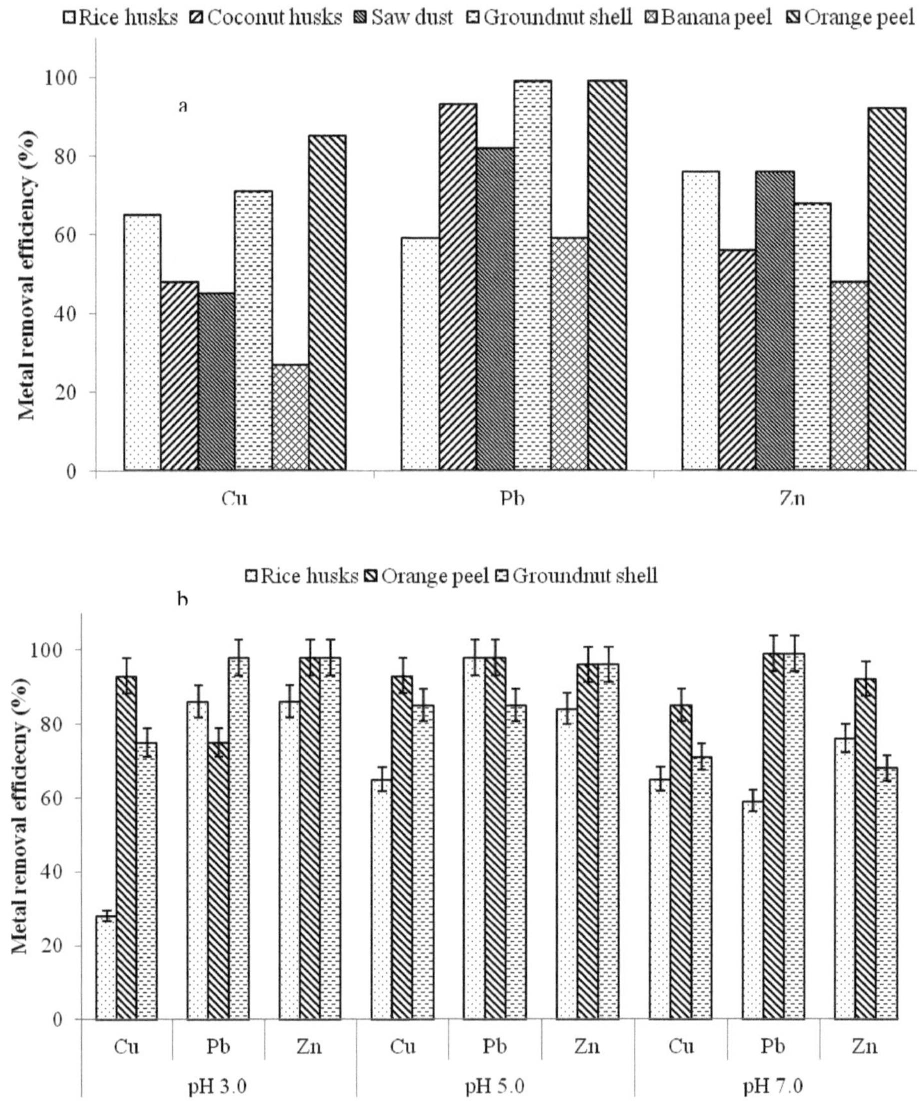

Figure 3.2 Screening experiments a) six adsorbents at pH 7.0 and b) three selected adsorbents at pH 3.0, 5.0 and 7.0

Leaching tests were also performed on the different adsorbents to cognize whether the adsorbents could pose post-treatment problems. The DOC content and pH of the leachate from each adsorbent is shown in the Table 3.3. The leachate of orange peel contained high DOC (106.1 mg/g) when compared to the other adsorbents. Therefore, orange peel was opted for further experiments and groundnut shell was selected as the main adsorbent for further optimization studies.

Table 3.3 The DOC content and final pH of the leachate from adsorbents (Initial pH = 7.0)

Adsorbents	DOC (mg/L)	Final pH
Orange peel	106.1	4.3
Banana peel	87.9	4.8
Coconut Husk	66.6	4.1
Groundnut shell	5.0	5.3
Rice Husks	3.5	6.2
Saw dust	13.2	6.3

3.3.2 Statistical analysis and model development

The analysis of variances (ANOVA) from the program MINITAB showed the coefficient of each individual parameter, interaction between the different parameters and constant value. The significant of the parameter was determined at level of $p < 0.5$. Majorly, all individual selected parameter had a big impact on metal uptake and removal efficiency. The interaction effect between pH×pH was one of the common features among all metal removal efficiency, while other interaction parameters differed from each others. According to the result obtained, the prediction equations were developed as shown below (equation 3.8 - 3.13).

$$\text{Cu removal efficiency (\%)} = 70.127 + 5.075\,(dose) + 5.897(pH) - 6.581\,(concentration) + 4.271\,(time) - 3.730\,(pH)^2 - 3.642\,(time)^2 \quad (3.8)$$

$$\text{Cu uptake (mg/g)} = 1.378 - 0.304\,(dose) + 0.128\,(pH) + 0.449\,(concentration) + 0.094\,(time) + 0.080\,(dose)^2 - 0.082\,(pH)^2 - 0.075\,(concentration)^2 - 0.080\,(time)^2 - 0.113\,(dose \times concentration) \quad (3.9)$$

$$\text{Pb removal efficiency (\%)} = 75.51 + 10.4\,(dose) + 5.85\,(time) + 5.78\,(pH) - 12.49\,(concentration) - 4.89\,(\text{pH})^2 \quad (3.10)$$

$$\text{Pb uptake (mg/g)} = 2.492 - 0.329\,(dose) + 0.197\,(time) + 0.206\,(pH) + 0.656\,(concentration) \quad (3.11)$$

$$\text{Zn removal efficiency (\%)} = 53.43 + 8.12\,(dose) + 13.19\,(pH) - 12.31(concentration) + 3.79\,(time) - 5.72\,(pH)^2 + 2.60\,(concentration)^2 - 2.24\,(time)^2 \quad (3.12)$$

$$\text{Zn uptake (mg/g)} = 1.049 - 0.112\,(dose) + 0.261\,(pH) + 0.228\,(concentration) + 0.073\,(time) - 0.121\,(pH)^2 - 0.093\,(dose \times pH) \quad (3.13)$$

3.3.2.1 Optimization model validation

Experiments were performed at the optimum conditions for Cu, Pb and Zn removal (Table 3.5) and random conditions (data not shown) in order to compare the prediction power of the model. The predicted and experimental values for all the experiments are shown in Figure 3.3 with their respective R^2 values. In general, the predicted values were a bit higher than results obtained from the experiment.

3.3.3 Main effects and interaction effect of parameters on metal uptake

Figure 3.4 shows the main effects of different process parameters on the metal uptake capacity. In case of Cu (Figure 3.4b), increasing the level of different parameters increased the metal uptake capacity, while the adsorbent dosage showed an opposite trend which were in the similar patterns to Pb and Zn (Figure 3.4d and f). In common, for all metal uptakes, the pH range which had a high uptake rate was between 2.0 and 4.0 and the contact time ranged between 10 to 70 min before the uptake capacity rate tends to become less effective. Moreover, the dosage of sorbent had less impact to the Zn uptake compared to Cu and Pb.

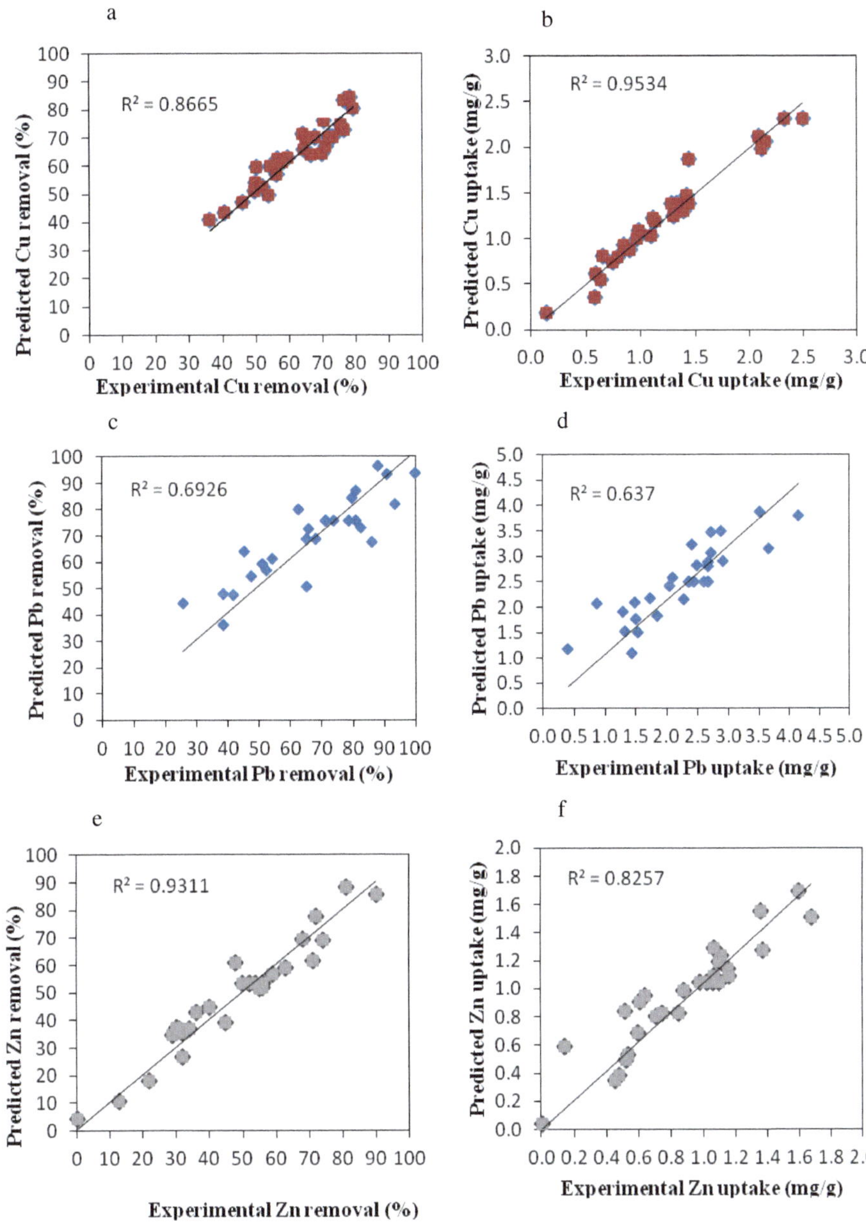

Figure 3.3 Relationship between predicted and experimental values of: a) Cu removal efficiency, b) Cu uptake c) Pb removal efficiency, d) Pb uptake, e) Zn removal efficiency and f) Zn uptake

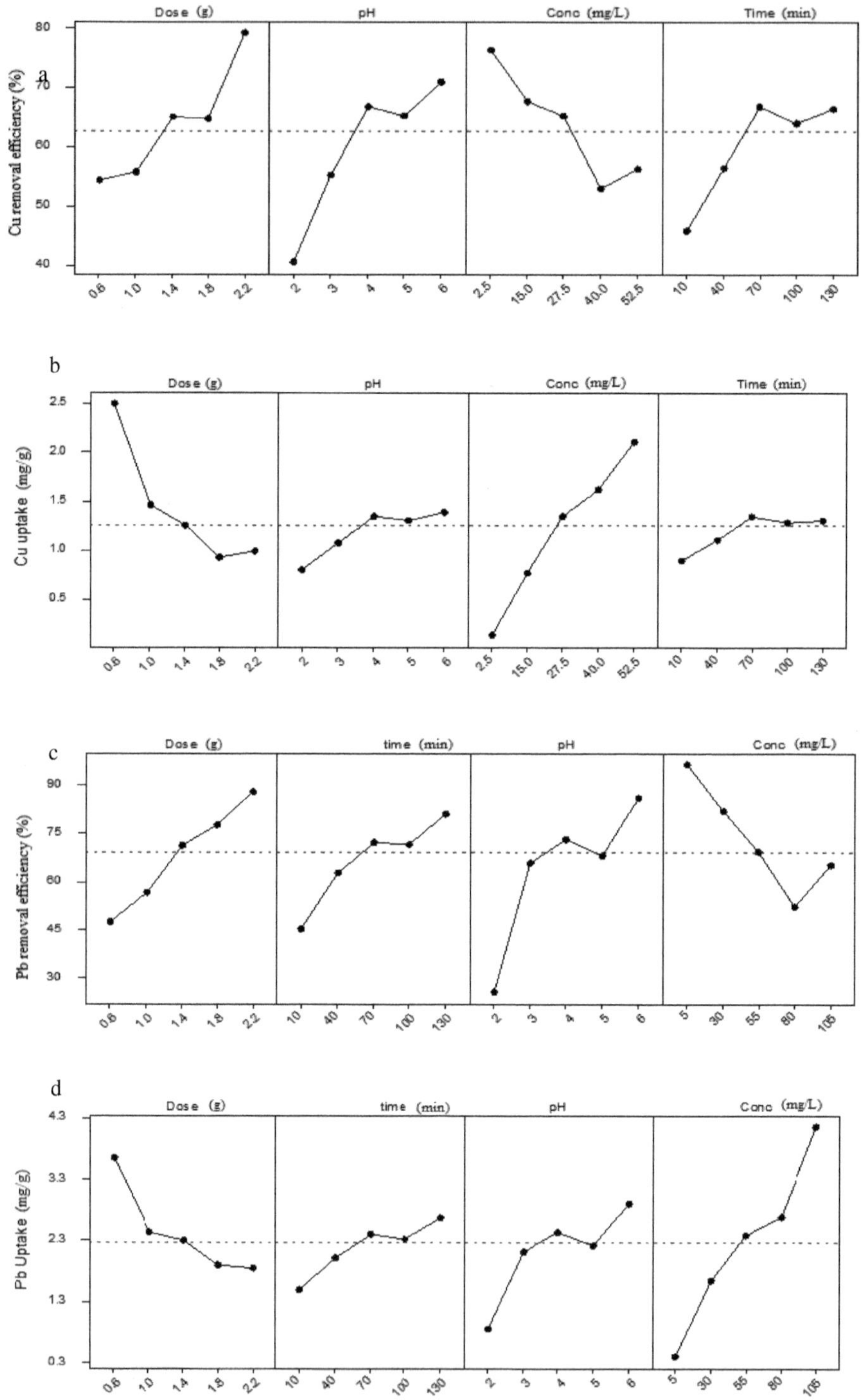
a
Dose (g)
pH
Conc (mg/L)
Time (min)
Cu removal efficiency (%)
b
Cu uptake (mg/g)
c
time (min)
Pb removal efficiency (%)
d
Pb Uptake (mg/g)

e

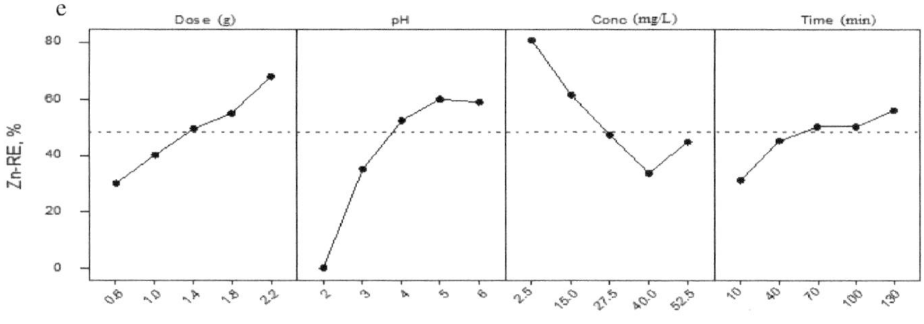

f

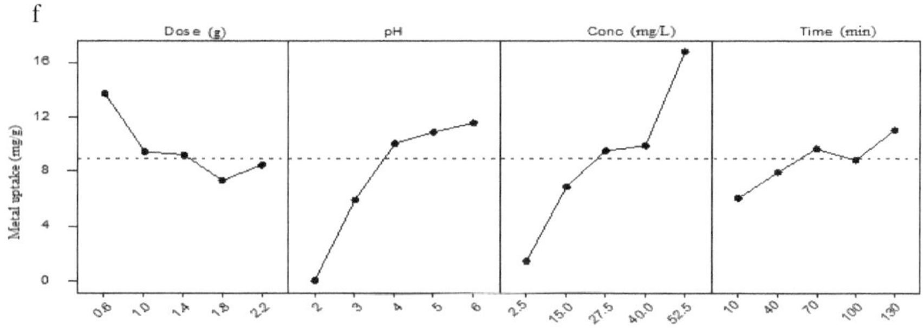

Figure 3.4 Main effects of process parameters a) Cu removal b) Cu uptake, c) Pb removal, d) Pb uptake, e) Zn removal and f) Zn uptake

For the interaction effects of parameters on metal uptake capacity, the experimental results were analyzed using surface and contour plots to elucidate the interaction effects between the different process parameters on the metal uptake capacity of the adsorbent. During each analysis, two parameters were kept constant, as indicated in these plots, while the other parameters were varied from low to high values. Figure 3.5 shows the interaction effects of parameters on the Cu, Pb and Zn uptake capacity by groundnut shell.

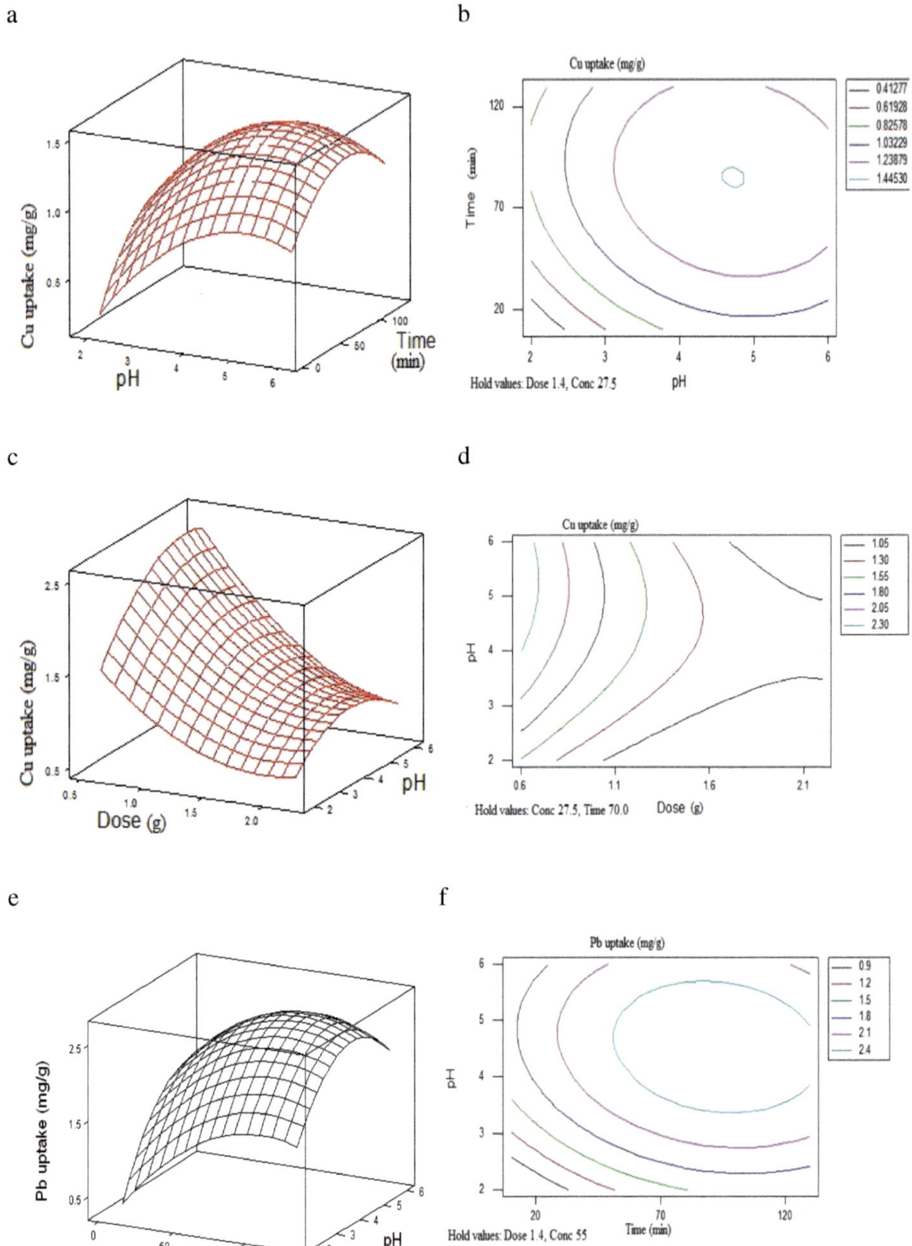
a
b
Cu uptake (mg/g)
Time (min)
pH
0.41277
0.61928
0.82578
1.03229
1.23879
1.44530
Hold values: Dose 1.4, Conc 27.5
c
d
Dose (g)
1.05
1.30
1.55
1.80
2.05
2.30
Hold values: Conc 27.5, Time 70.0
e
f
Pb uptake (mg/g)
Time (min)
0.9
1.2
1.5
1.8
2.1
2.4
Hold values: Dose 1.4, Conc 55

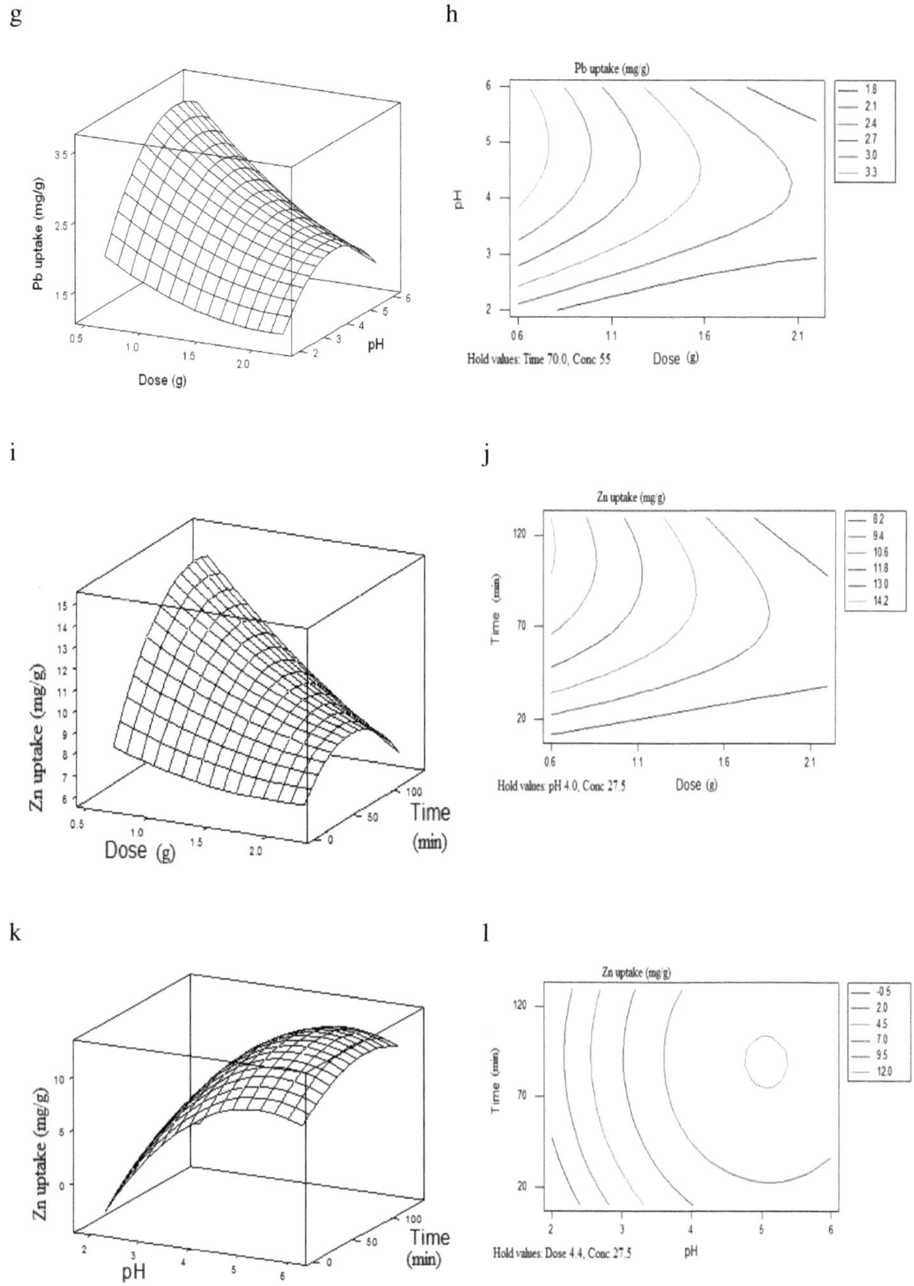

Figure 3.5 Surface plots and their corresponding contour plots for Cu (a-d), Pb (e-h) and Zn (i-l) uptake

Figure 3.5 (a - d) shows the interaction effect between pH×time and dose×pH on Cu uptake capacity. It can be clearly seen that increasing the contact time and pH, the highest Cu uptake (1.44 mg/g) was obtained around pH 4.8 and 80 min which thereafter started to decline. For Pb uptake capacity (Figure 3.5e - h), an increase in pH and time (dose and initial concentration were constant) also gave similar trend to Cu uptake in which maximal uptake of 2.4 mg/g was obtained at pH around 4.5 and 90 min. However in this study, the maximum Pb uptake while varied pH and dose showed the similar pattern of reaching maximum uptake capacity before their values started to decline. However, the values of the maximum condition were in the lowest limit of the concerned parameter. Likewise, the Zn uptake capacity also gave a similar pattern as Cu and Pb for the interaction between pH and time with a maximal uptake capacity of 12 mg/g, at a pH of 5.0 and 100 min contact time. For the effect of dose×time on Zn uptake capacity, the highest uptake (14.2 mg/g) was in the range of dose lower than 0.6 g with the contact time ~ 100 min.

3.3.4 Regeneration of the adsorbent

In this study, adsorption/desorption experiments were performed in six cycles. Among the five desorbing solutions, HCl showed the highest efficiency in desorbing the metals ions followed by Na_2EDTA, while the least effective regenerating agent was sodium acetate (Figure 3.6). The amount of metal desorbed using HCl after the first adsorption cycle was 0.24, 0.38 and 0.13 mg/g for Cu, Pb and Zn, respectively, which were 92.3, 95.0 and 100.0% of the adsorbed metal. Hence, HCl was selected for further adsorption/desorption cycle tests, at the respective optimal conditions obtained for the individual metals. The amount of metal adsorbed/desorbed by the adsorbent showed a near similar trend in the profiles for all the HMs (Figure 3.7). For the first cycle, the amounts of Cu, Pb and Zn adsorbed on groundnut shell were 1.02, 4.90 and 0.71 mg/g, while the amount desorbed were 0.87, 4.30 and 0.71 mg/g (85.1, 89.2 and 100.0% of adsorption), respectively. Moreover, both adsorption and desorption of the metals on groundnut shell decreased and became almost constant after three cycles of operation.

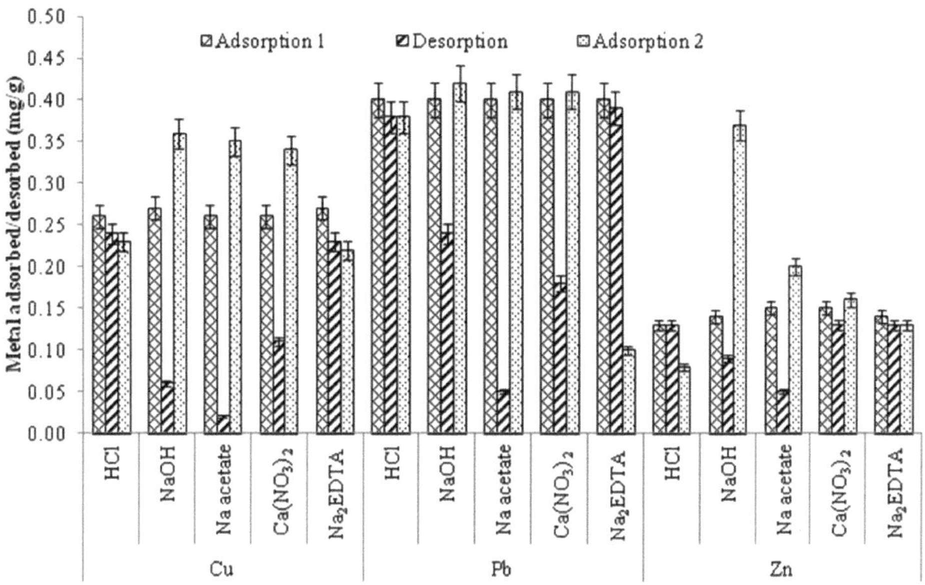

Figure 3.6 Screening of desorption solution for Cu, Pb and Zn on groundnut shell

3.4 Discussion

3.4.1 Effects of process parameters on the adsorption efficiency and metal uptake capcity

In this study, the parameters which affected the metal removal efficiency were pH, metal concentration, contact time and adsorbent dosage. There are some report regarding the effect of pH on metal speciation, precipitation and the activeness of the binding site on the adsorbent. Equation 3.14 shows a typical example of HMs adsorption using organic adsorbents (M is the metal ion, x is the charge, A is the adsorbent active site, and H^+ is the hydrogen ion concentration). Bayramoğlu and Arıca (2008) also reported that heavy metal such as Cu, Pb and Zn binds on organic adsorbents, which is strongly pH dependent.

$$M^{x+} + AH_x \rightleftharpoons AM + xH^+ \tag{3.14}$$

At low pH, where the concentration of H^+ ions are relatively high, the competition with metal ions for binding sites on the surface of the adsorbent also increases resulting in low metal uptake capacity (Zhang, 2011).

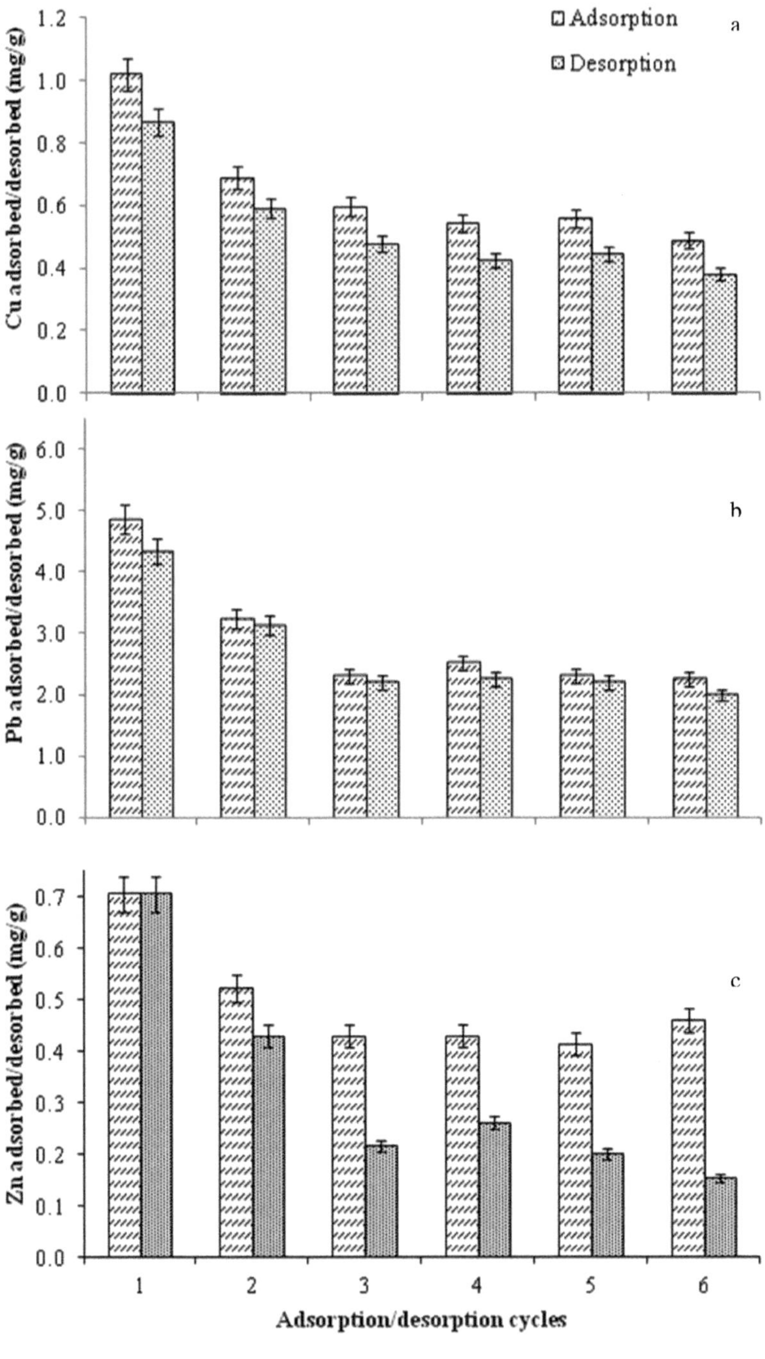

Figure 3.7 Absorption/desorption cycles of a) Cu, b) Pb and c) Zn

As pH increases, the metal uptake capacity also increases due to the presence of negatively charged functional groups on the adsorbent surface, resulting in increased binding sites. In order to overcome the competition for the active binding site, a change in the adsorbent dosage or initial metal concentration play a major role. According to Figure 3.4, it can be interpreted that increasing the adsorbent dosage increases the removal efficiency, while it decreases the metal uptake capacity. This scenario implies that the increasing number of active sites help the metal ions to have more chance to bind to the active sites while, it can also alter with each other to attract metal ions on to the material (Ahmad et al., 2009; Lasheen et al., 2012). Acheampong et al. (2010) and Yu et al. (2000) also reported the similar observations in their experimental studies. Another important parameter is contact time which mainly involves with the saturation of the active sites on adsorbent surface. The reaction between the active site and metal ions binding are relatively fast during the intake period of metals exposure. Thereafter, the active sites are getting occupied by metal ions which resulted in the remaining ions compete among each other for the free active sites (Amuda et al., 2007). A similar trend was observed in this study, when the contact time increased, the removal efficiency also increased up to a certain level before it started to remain constant or increase at a rather slower rate.

3.4.2 Adsorbent selection and technology application

In this study, groundnut shell was selected as the most suitable adsorbent, although orange peel outperformed other adsorbents. Agricultural wastes have hydroxyl groups which can act as binding sites for HMs. The tricarboxyl groups which can be found in the citrus fruit like orange can form organic ligands that can bind metals ions. This makes its performance higher than other adsorbents (Kamsonlian et al., 2011). Schiewer and Iqbal (2010) also found that the pectin contents in the citrus fruit majorly contributed to the better metal adsorption, especially Cd in their study. However, orange peel released higher DOC than groundnut shell, which in turn decreased the final pH of the effluent (Table 3.4). This will cause further burden on the treatments before discharge into natural water body (Chen et al., 2010; Shukla and Pai, 2005; Suantak Kamsonlian, 2011; Wan Ngah and Hanafiah, 2008). As a consequence, the groundnut shell was found to be more advantageous in this criterion and was selected.

From the result obtained from regeneration experiments (Figure 3.5 and 3.6), the efficiency of groundnut shell decreased with an increase in the number of regeneration cycles. One of the explanations toward this scenario is that the strongly bound metal ions to the adsorbent surface which cannot be longer replace by the ions from desorbing solution (Chen et al., 2010). The results obtained from this study show that the maximum number of adsorption/desorption cycles for groundnut shell was 3 cycles. However during screening test with different reagents, some reagents made the groundnut shell exhibit higher adsorbing efficiency in the second cycle, which might resulted from the surface modification. NaOH is one of the well-known reagents which has the ability to modify the surface properties of many adsorbents (Shim et al., 2001). Acheampong et al. (2013) also reported that HCl was one of the most efficient desorbing solutions in which concentration of the reagents play an important role in the desorbed efficiency.

In term of real application, adsorbent characteristics, technical applicability, type of modification, cost-effectiveness and initial metal concentration play a major role in determining the suitability of an adsorbent. Recently, several studies had been carried out using various modified adsorbents (Dubey and Gopal, 2007; Feng and Guo, 2012; Feng et al., 2009; Lasheen et al., 2012; Li et al., 2007; Noeline et al., 2005; Shukla and Pai, 2005; Wong et al., 2003) and it can be interpreted that the modification of adsorbents can improve the removal capacity of the adsorbent (Kurniawan et al., 2006). However, for practical reasons (ease of usage and economically attractive), it will be better if an adsorbent can be efficient in its raw state or with minimal pre-treatment. Moreover, in terms of technological applications, groundnut shell may not be a cost-effective material for HMs removal from wastewater if the material is not available locally. The materials tested in this study were limited to six materials. There are many options available locally which may yield higher efficiency for other metals or locally cost-effective than groundnut shell. Agricultural and food industrial wastes, tree barks, petroleum wastes, fly ash, clays and mud are examples of feasible sorbent matrices. In addition, this study mainly focuses on optimizing/screening in order to estimate the conditions and materials for metal contaminated wastewater treatment using statistically based batch experiments. The degradation of the material, their longevity in column leaching, material preparation for scale up, regeneration of the material and interaction between different metals should be further studied. Acheampong and Lens (2014) have proved that the concept of using low-cost adsorbent technology can be applicable in large scale, where in a pilot plant system was installed for the gold mining effluent, and their result showed satisfactory level treatment. Volesky (2003; 2007) also demonstrated that low-cost adsorbent is one of the most cost-effective technology for metal contaminated wastewater treatment.

3.5 Conclusions

The results from this study showed that groundnut shell was found to be a suitable low cost adsorbent for the removal of Cu, Pb and Zn with a maximum removal efficiency of 75% (pH 5.0 - 7.0), 98 % (pH 5.0) and 86% (pH 3.0), respectively. Moreover, the material was regenerated successfully with 0.2 M HCl upto three cycles. Response surface methodology can be used as an effective tool to optimize groundnut shell adsorption capacity for HMs. Individual factors such as initial metal concentration, adsorbent dosage, pH and contact time play an important role on the metal adsorption capacity of groundnut shell, however, only interaction between time×initial concentration, pH×initial concentration and pH×time were significantly affected the absorption capacity.

References

Acheampong, M.A., Dapcic, A.D., Yeh, D. and Lens, P.N., 2013. Cyclic sorption and desorption of Cu (II) onto coconut shell and iron oxide coated sand. Sep. Sci. Technol., 48(18): 2786-2794.

Acheampong, M.A. and Lens, P.N., 2014. Treatment of gold mining effluent in pilot fixed bed sorption system. Hydrometallurgy, 141: 1-7.

Acheampong, M.A., Meulepas, R.J.W. and Lens, P.N.L., 2010. Removal of heavy metals and cyanide from gold mine wastewater. J. Chem. Technol. Biotechnol., 85(5): 590-613.

Ahmad, A. et al., 2009. Removal of Cu (II) and Pb (II) ions from aqueous solutions by adsorption on sawdust of Meranti wood. Desalt., 247(1): 636-646.

Alcolea, A. et al., 2012. Heavy metal removal of intermittent acid mine drainage with an open limestone channel. Miner. Eng., 26(0): 86-98.

Alslaibi, T.M., Abustan, I., Ahmad, M.A. and Foul, A.A., 2013. Application of response surface methodology (RSM) for optimization of Cu^{2+}, Cd^{2+}, Ni^{2+}, Pb^{2+}, Fe^{2+}, and Zn^{2+} removal from aqueous solution using microwaved olive stone activated carbon. J. Chem. Technol. Biotechnol., 88(12): 2141-2151.

Amini, M. et al., 2008. Application of response surface methodology for optimization of lead biosorption in an aqueous solution by *Aspergillus niger*. J. Hazard. Mater., 154(1-3): 694-702.

Amuda, O.S., Giwa, A.A. and Bello, I.A., 2007. Removal of heavy metal from industrial wastewater using modified activated coconut shell carbon. Biochem. Eng. J., 36(2): 174-181.

Anupam, K., Dutta, S., Bhattacharjee, C. and Datta, S., 2011. Adsorptive removal of chromium (VI) from aqueous solution over powdered activated carbon: Optimisation through response surface methodology. Chem. Eng. J., 173(1): 135-143.

Barrera, H., Ureña-Núñez, F., Bilyeu, B. and Barrera-Díaz, C., 2006. Removal of chromium and toxic ions present in mine drainage by Ectodermis of *Opuntia*. J. Hazard. Mater., 136(3): 846-853.

Bayramoğlu, G. and Arıca, M.Y., 2008. Removal of heavy mercury (II), cadmium (II) and zinc (II) metal ions by live and heat inactivated *Lentinus edodes* pellets. Chem. Eng. J., 143(1): 133-140.

Bhatia, D., Kumar, R., Singh, R., Chadetrik, R. and Bishnoi, N.R., 2011. Statistical modelling and optimization of substrate composition for bacterial growth and cadmium removal using response surface methodology. Eco. Eng., 37(12): 2076-2081.

Chen, H. et al., 2010. Removal of copper (II) ions by a biosorbent - *Cinnamomum camphora* leaves powder. J. Hazard. Mater., 177(1-3): 228-236.

Chuah, T.G., Jumasiah, A., Azni, I., Katayon, S. and Thomas Choong, S.Y., 2005. Rice husk as a potentially low-cost biosorbent for heavy metal and dye removal: an overview. Desalination, 175(3): 305-316.

Dakiky, M., Khamis, M., Manassra, A. and Mer'eb, M., 2002. Selective adsorption of chromium (VI) in industrial wastewater using low-cost abundantly available adsorbents. Adv. Environ. Res., 6(4): 533-540.

Davies, H., Weber, P., Lindsay, P., Craw, D. and Pope, J., 2011. Characterisation of acid mine drainage in a high rainfall mountain environment, New Zealand. Sci. Total Environ., 409(15): 2971-2980.

Dubey, S.P. and Gopal, K., 2007. Adsorption of chromium(VI) on low cost adsorbents derived from agricultural waste material: A comparative study. J. Hazard. Mater., 145(3): 465-470.

Eaton, A.D., Franson, M.A.H., Association, A.P.H., Association, A.W.W. and Federation, W.E., 2005. *Standard Methods for the Examination of Water & Wastewater*. American Public Health Association, Washington D.C., the United States.

Feng, N.-C. and Guo, X.-Y., 2012. Characterization of adsorptive capacity and mechanisms on adsorption of copper, lead and zinc by modified orange peel. Trans. Nonferrous Met. Soc. of China, 22(5): 1224-1231.

Feng, N., Guo, X. and Liang, S., 2009. Adsorption study of copper (II) by chemically modified orange peel. J. Hazard. Mater., 164(2-3): 1286-1292.

Fu, F. and Wang, Q., 2011. Removal of heavy metal ions from wastewaters: a review. J. Environ.Manage., 92(3): 407-418.

Hashim, M., Mukhopadhyay, S., Sahu, J.N. and Sengupta, B., 2011. Remediation technologies for heavy metal contaminated groundwater. J. Environ. Manage., 92(10): 2355-2388.

Kalantari, K. et al., 2014. Rapid adsorption of heavy metals by Fe_3O_4/talc nanocomposite and optimization study using response surface methodology. Int. J. Mol. Sci., 15(7): 12913-12927.

Kamsonlian, S., Suresh, S., Majumder, C. and Chand, S., 2011. Characterization of banana and orange peels: biosorption mechanism. Int. J. Sci. Technol. Manage, 2(4): 1-7.

Khan, N.A., Ibrahim, S. and Subramaniam, P., 2004. Elimination of heavy metals from wastewater using agricultural wastes as adsorbents. Malaysian J. Sci., 23(1): 43-51.

Kiran, B. and Thanasekaran, K., 2011. Copper biosorption on *Lyngbya putealis*: application of response surface methodology (RSM). Int. Biodeterior. Biodegradation, 65(6): 840-845.

Kurniawan, T.A., Chan, G.Y.S., Lo, W.-H. and Babel, S., 2006. Comparisons of low-cost adsorbents for treating wastewaters laden with heavy metals. Sci. Total Environ., 366(2-3): 409-426.

Landaburu-Aguirre, J., Pongrácz, E., Perämäki, P. and Keiski, R.L., 2010. Micellar-enhanced ultrafiltration for the removal of cadmium and zinc: use of response surface methodology to improve understanding of process performance and optimisation. J. Hazard. Mater., 180(1): 524-534.

Lasheen, M.R., Ammar, N.S. and Ibrahim, H.S., 2012. Adsorption/desorption of Cd (II), Cu (II) and Pb (II) using chemically modified orange peel: equilibrium and kinetic studies. Solid State Sci., 14(2): 202-210.

Li, Q., Zhai, J., Zhang, W., Wang, M. and Zhou, J., 2007. Kinetic studies of adsorption of Pb (II), Cr (III) and Cu (II) from aqueous solution by sawdust and modified peanut husk. J. Hazard. Mater., 141(1): 163-167.

Liang, S., Guo, X.-Y., Feng, N.-C. and Tian, Q.-H., 2010. Effective removal of heavy metals from aqueous solutions by orange peel xanthate. Trans. Nonferrous Met. Soc. China, 20, Supplement 1(0): s187-s191.

Noeline, B.F., Manohar, D.M. and Anirudhan, T.S., 2005. Kinetic and equilibrium modelling of lead(II) sorption from water and wastewater by polymerized banana stem in a batch reactor. Sep. Purif. Technol., 45(2): 131-140.

Schiewer, S. and Iqbal, M., 2010. The role of pectin in Cd binding by orange peel biosorbents: a comparison of peels, depectinated peels and pectic acid. J. Hazard. Mater., 177(1-3): 899-907.

Shim, J.-W., Park, S.-J. and Ryu, S.-K., 2001. Effect of modification with HNO_3 and NaOH on metal adsorption by pitch-based activated carbon fibers. Carbon, 39(11): 1635-1642.

Shukla, S. and Pai, R.S., 2005. Adsorption of Cu (II), Ni (II) and Zn (II) on dye loaded groundnut shells and sawdust. Sep. Purif. Technol., 43(1): 1-8.

Suantak Kamsonlian, S.S., C.B. Majumder and S. Chand, 2011. Characterization of banana and orange peel: Biosorption mechanism. Int. J. Sci. Technol. Manage., 2(4).

Sud, D., Mahajan, G. and Kaur, M., 2008. Agricultural waste material as potential adsorbent for sequestering heavy metal ions from aqueous solutions: a review. Bioresour. Technol., 99(14): 6017-6027.

Turan, N.G., Elevli, S. and Mesci, B., 2011. Adsorption of copper and zinc ions on illite: Determination of the optimal conditions by the statistical design of experiments. Appl. Clay Sci., 52(4): 392-399.

Volesky, B., 2003. Sorption and biosorption. BV Sorbex.

Volesky, B., 2007. Biosorption and me. Water Res., 41(18): 4017-4029.

Wan Ngah, W.S. and Hanafiah, M.A.K.M., 2008. Removal of heavy metal ions from wastewater by chemically modified plant wastes as adsorbents: a review. Bioresource Technol., 99(10): 3935-3948.

Wang, S. and Peng, Y., 2010. Natural zeolites as effective adsorbents in water and wastewater treatment. Chem. Eng. J., 156(1): 11-24.

Wong, K.K., Lee, C.K., Low, K.S. and Haron, M.J., 2003. Removal of Cu and Pb from electroplating wastewater using tartaric acid modified rice husk. Process Biochemistry, 39(4): 437-445.

Yu, B., Zhang, Y., Shukla, A., Shukla, S.S. and Dorris, K.L., 2000. The removal of heavy metal from aqueous solutions by sawdust adsorption — removal of copper. J. Hazard. Mater., 80(1-3): 33-42.

Zhang, M., 2011. Adsorption study of Pb (II), Cu (II) and Zn (II) from simulated acid mine drainage using dairy manure compost. Chem. Eng. J., 172(1): 361-368.

Chapter 4 The Performance of Sulfate Reducing Bacteria Biotechnology in an Inverse Fluidized Bed Bioreactor

The modified version of this chapter was published as:

Janyasuthiwong S., Rene E.R., Esposito G. and Lens P.N.L. (2015). Effect of pH on the performance of sulfate and thiosulfate fed sulfate reducing inverse fluidized bed reactors. *J. Environ. Eng., In press.*

DOI: 10.1061/(ASCE)EE.1943-7870.0001004

Summary

Sulfate reducing bacteria (SRB) based technologies have gained a lot of attention in the field of wastewater treatment, especially to treat metal contaminated wastewaters. An inverse fluidized bed (IFB) bioreactor is a versatile bioreactor configuration that uses SRB technology for metal removal and recovery from wastewater. Apart from sulfate which is commonly used as an electron acceptor, thiosulfate is another potential candidate for this process. In this study, the performance of two IFB bioreactors which were operated at pH 7.0 (R1) and 5.0 (R2) using sulfate and thiosulfate as the electron acceptors were evaluated. The electron donor used in this study was ethanol and the COD/electron acceptor ratio (SO_4^{2-} or $S_2O_3^{2-}$) was kept constant at 1.0. By using sulfate as the electron acceptor, the average COD removal efficiency was 75.0 and 58.0% at pH 7.0 and 5.0, respectively, while the sulfate removal efficiency was 74.4 and 50.4%, respectively. The average sulfide production was 246.3 and 150.7 mg/L at pH 7.0 and pH 5.0, respectively. Using thiosulfate as the electron acceptor, slightly higher sulfate reduction activities were achieved when compared to sulfate at pH 5.0. The maximum COD removal efficiency was 54.8% and 162.7 mg/L sulfide was produced. The COD and sulfate removal efficiencies as well as the total sulfide production profiles in the IFB reactor fed with sulfate was modeled using a three layered artificial neural networks (ANN). The results showed that the developed ANN model with a topology of 3-7-3 was able to give good predictions of the performance variables. Moreover, the sensitivity analysis from ANN showed that this process is mainly pH dependant.

4.1 Introduction

The previous chapter (chapter 3) has shown one of the cost effective technology for metal contaminated wastewater treatment by using agricultural waste for adsorption. In this chapter, another technology which use microbiological reactor for metal removal and recovery will be discuss in which sulfate reducing bacteria (SRB) is one of the well-known methods. Some wastewaters contain high sulfate concentrations like acid mine drainage, pulp and paper wastewater and construction waste leachate. These generate problems directly, e.g., acid rain, or indirectly after the reduction of sulfate to H_2S, e.g., volatile acid gases and corrosion of materials especially in sewer systems in the cities (Lens et al., 2001; Lens et al., 2000). There are many methods available for the treatment of these wastewaters like membrane separation and precipitation, but one of the most cost effective techniques for industrial facilities is the biological treatment process using SRB (Liang et al., 2013). In this process, the effluent can be further used for other processes, like the treatment of metal contaminated wastewater (Jong et al., 2003).

SRB play important roles in many elemental cycles apart from the sulfur cycle (Heidelberg et al., 2004). Sulfate reducing (SR) activities occur under anaerobic conditions, wherein sulfate (SO_4^{2-}) is reduced to sulfide (H_2S, HS^- and S^{2-}). The terminal electron acceptor in SR is in most cases sulfate. However, SRB can also use other electron acceptors like thiosulfate ($S_2O_3^{2-}$). Several waste streams from photograph film processing, metallurgic activities and leaching of sulfide minerals like pyrite (FeS_2) generate thiosulfate rich wastewater (Liljeqvist et al., 2011; Schippers et al., 1999). The incentive to use or treat thiosulfate is that it can also cause similar environmental problems as sulfate discharged without adequate treatment (González - Sánchez et al., 2008). Thiosulfate oxidation depletes dissolved oxygen and decreases the pH through sulfuric acid formation (Equation 4.1):

$$S_2O_3^{2-} + 2\ O_2 + H_2O \rightarrow 2\ SO_4^{2-} + 2\ H^+ \qquad (4.1)$$

The pH is considered to be one of the key factors which govern many metabolic mechanisms, such as homeostasis, channeling for compound transportation and dissociation of electron donor. The favorable pH for SR is between 5.0 and 9.0 and no activities were found at a pH below 2.0 (Jong et al., 2006; Neculita et al., 2007). However, there are some SRB species that can tolerate harsh conditions like those usually observed in acid mine drainage which has a low pH and high metal concentrations. Moreover, the sulfide speciation is also pH dependent, which will determine in which forms (H_2S, HS^- and S^{2-}) the product will be present (Sawyer et al., 2003). The electron donor is another factor which drives the activity of SRB. There are many options available, varying from single compounds to substrate mixtures including organic wastewaters (Cao et al., 2012; Liamleam et al., 2007a). Ethanol is one of the cost effective electron donors considered by the price as well as ease of handling and use (Liamleam et al., 2007a). However, the major drawbacks of using ethanol are the low growth rate of SRB and the acetate generation which can cause toxicity to the SRB and residual COD pollution in the effluent (Nagpal et al., 2000; Stucki et al., 1993).

Biological wastewater treatment systems are time variant and their performance is often dictated by the micro and macro kinetics of the microbe-mediated processes. Models developed to represent the behavior of such systems should be reliable and versatile to provide meaningful predictions of the state variables. An artificial neural network (ANN) is one of the tools used for predicting the relationship of each input parameter with a given set of outputs which has drawn the attention of many researchers in the field of natural, engineering and biological sciences. In ANNs, a set of numerical data containing the input and output variables can be mapped adequately by a suitable algorithm and be used for system identification, prediction and pattern recognition purposes (Li et al., 2009; Mjalli et al., 2005). There are many studies which use ANN for many kinds of wastewater treatment, for example penicillin-G (Mullai et al., 2011), tannery (Ongen et al., 2013), and high sulfate and metal contaminated (Atasoy et al., 2013; Bayrakdar et al., 2009; Sahinkaya et al., 2010) wastewater treatment. These studies have clearly demonstrated the efficiency of neural networks. The aim of this study was to evaluate the sulfide production using sulfate and thiosulfate as electron acceptors in two inverse fluidized bed (IFB) reactors operated in parallel and study the effects of pH on sulfate reducing activities. The performance of the IFB operated with sulfate as the electron acceptor was modeled using a three-layer back propagation neural network using pH, the sulfate and the COD concentration as the input variables.

4.2 Material and methods

4.2.1 Medium composition

The synthetic medium used in both the batch experiments and continuous reactors consisted of NH_4Cl 300 mg/L, $MgCl_2 \cdot 6H_2O$ 120 mg/L, $KHPO_4$ 200 mg/L, KCl 250 mg/L, $CaCl_2 \cdot 2H_2O$ 15 mg/L and yeast extract 20 mg/L, respectively. The electron acceptor sources used were Na_2SO_4 and $Na_2S_2O_3 \cdot 5H_2O$, while ethanol was used as electron donor at a COD to sulfate/thiosulfate ratio of 1.0. The initial pH of the synthetic medium was 7.0 ± 0.2, adjusted using 1.0 M NaOH and 1.0 M HCl to the set condition. All chemicals used in this study were of analytical grade.

4.2.2 Source of the biomass

The carrier material for the biomass was low-density polyethylene (PE) beads (Purell Pe 1810E, Basell Polyolifins, the Netherlands) with a diameter of 3 mm. The surface of the carrier materials was roughened before starting the experiment with sand abrasion and rinsing the sand from the beads. The total volume of the carrier material was 600 mL with the expansion of 1/3 of the column height. The biomass used for the experiments was obtained from Biothane Systems International (Delft, the Netherlands). The biomass was kept in close loop recirculation for 2 days in order to enhance the attachment of biomass onto the carrier materials. The volatile suspended solid (VSS) concentration of the seed sludge was 39.8 g/L of the mixed liquid sample (wet weight basis).

4.2.3 Batch experiments

Sulfate reducing activity tests were done in triplicates in 115 mL bottles fitted with air-tight septum cap, at room temperature and placed on an orbital shaker at 150 rpm as described by Villa-Gomez et al. (2011). The bottles were filled with 5 mL biomass (from the seed sludge or IFB bioreactor) and 100 mL synthetic medium. During batch experiments, sulfate reducing activities were compared for two types of biomass, *viz.*, sulfate reducing granular sludge and biomass attached to the PE beads from the IFB, using sulfate and thiosulfate as electron acceptors. Experiments were performed at initial pH values of 3.0, 4.0, 5.0, 6.0 and 7.0. Batch incubation lasted 28 days for both experiments. The samples were collected on day 1, 3, 7, 14, 21 and 28 to analyze for residual COD, sulfide and sulfate concentrations, while the pH was measured at the end of the experiment using a 691 Metrohm pH meter and a SenTix 21 WTW pH electrode.

After the IFB bioreactor operation at pH 5.0 period Sulfate I failed, acetate toxicity tests were done at pH 5.0 and 7.0 with initial acetate concentrations ranging from 100.0 to 500.0 mg/L based on the residual acetate and COD concentrations found during the last week of the Sulfate I period at pH 5.0 to evaluate toxic effects of acetate on the SR performance.

4.2.4 Inversed fluidized bed reactor set up

The reactor column (1.0 m height, diameter 0.08 m) had a total volume of 2.5 L. The reactors were constructed from a PMMA (polymethyl methacrylate) pipe mounted with removable caps at the top which consists of channels for gas outlet and recirculation flow (Figure 4.1). The influent was circulated *via* multichannel peristaltic pumps (Watson-Marlow BV, the Netherlands). The recirculation flow to mix the liquid inside the reactor and maintain the expansion of the carrier material was done by a magnetic driven pump model IWAKI MD-20R-22ON (Iwaki Holland BV, the Netherlands). The expansion of the carrier material zone was around 30% of the total volume and the equalization tank was connected to each reactor to maintain the level of liquid inside the column.

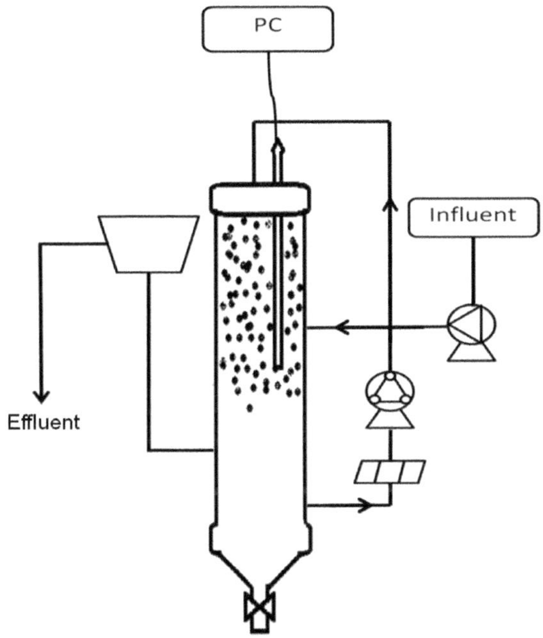

Figure 4.1 Schematic of the IFB reactor configuration used in this study

4.2.4.1 Reactor operation

Two reactors, R1 and R2, were operated with the same feeding conditions, except that the reactor pH in R1 was maintained at pH 7. 0 ± 0.2, while the pH of R2 was gradually decreased from pH 7.0 to 5.0 (Table 1). The start-up period of both reactors were conducted in a similar way to a previously operated IFB, as described by Villa-Gomez et al. (2011), except that a two day recirculation of the sludge was provided instead of one day during start up. The two reactors were operated at room temperatures (25.0 ± 2.0 °C). The hydraulic retention time was maintained at 24 hours throughout the study. The operating conditions were divided mainly into 3 periods, which were annotated Sulfate I, Sulfate II and Thiosulfate according to Table 4.1.

Table 4.1 Operating conditions during the different periods and phases applied to the reactors

Period	Experimental phases (days)		
Sulfate I	pH7.0	pH 6.0	pH 5.0
R1	0 - 124	-	-
R2	0 - 32	33 - 62	63 - 124
Recovery			
R1	124 - 158	-	-
R2	124 - 158	-	-
Sulfate II			
R1	158 - 252	-	-
R2	158 - 190	191 - 221	222 - 252
Thiosulfate			
R1	255 - 358	-	-
R2	255 - 287	287 - 318	319 - 358

4.2.5 Analytical methods

The samples from the reactors were collected once every 2 days and analyzed for residual COD, sulfide, sulfate (Sulfate I, Sulfate II and Thiosulfate periods) and thiosulfate concentrations (Thiosulfate period). For thiosulfate, extra samples were also taken from the influent tank. Temperature in the room was monitored daily which was averagely 25.0 ± 2.0 °C. The pH inside the reactor was monitored online using sulfide resistant pH electrodes (Prosense, Oosterhout, the Netrherlands) attached to an acquisition card (NI cDAQ-9174, National Instruments, the Netherlands) and controlled by the program LabVIEW 2009. COD was determined with the closed reflux method (APHA, 1999). The sulfide concentration was measured by a colorimetric method according to the procedure outlined by Cord-Ruwisch (1985). The absorbance of both COD and sulfide were measured with a spectrophotometer (PerkinElmer Lambda 20) at a wavelength of 600.0 and 480.0 nm, respectively.

For sulfate measurements, the sample was filtered through a 0.45 μm nitrocellulose filter (Millipore) and measured with an ion-chromatograph (ICS-1000 Dionex with ASI-100 Dionex) fitted with a IonPac AS14n column. The following conditions were used: sample injection volume - 10 μL, flow rate of Na_2CO_3/$NaHCO_3$ (8 mM/1mM) used as an eluent -

0.5 mL/min, temperature - 35°C, current - 35 mA and retention time - 8 minutes. Thiosulfate was determined by the iodometric method (APHA, 1999). Acetate was measured by a Varion 430 gas chromatography (GC) after acidification with 5% concentrated formic acid (Eaton et al., 2005).

4.2.6 Artificial neural network modeling

An artificial neural network based model was used as a tool to access the performance of reactor R2 during the periods Sulfate I and II. A three-layered neural network employing the standard back propagation algorithm with gradient was used to develop the ANN model for this study (Sahinkaya, 2009). As shown in Figure 2, the input parameters to the ANN model were operational pH (X_1), influent sulfate (X_2) and COD (X_3) concentrations (mg/L), while the output parameters were COD removal (Y_1, %), sulfate removal (Y_2, %) and sulfide production (Y_3, mg/L), respectively. ANN modeling was carried out using the shareware version of the neural network based multivariable statistical modeling software, NNMODEL (PA, USA).

A total of 99 data points from the periods Sulfate I and II of IFB operation (excluding the 29 days IFB recovery period) were divided into training (N_{Tr}=69) and test (N_{Te}=30) data sets. The data points for the training set were chosen in such a way to include the minimum and maximum values of the state variables that were representative of all the pH conditions tested in this study during these two periods. The data were normalized and scaled in the range 0 to 1 as described in Equation 4.2. A sigmoid transfer function was used in the hidden layer, while a linear transfer function was used in the output layer:

$$\hat{X} = \frac{X - X_{min}}{X_{max} - X_{min}} \qquad (4.2)$$

where, $\hat{X}$ is the normalized value, X_{min} and X_{max} are the minimum and maximum values of X, respectively.

The internal parameters of network, number of neurons in the hidden layer (N_H), learning rate (η), momentum term (μ), training count (T_c) and training vector size (ε) were chosen based on a trial and error approach (Rene et al. 2006). This was done by performing simulations at different initial settings of these parameters, by keeping one or two parameters at their default values, and by varying the other parameters slowly from low to high values. During network training, the value of ε was kept constant at 50. The best values of these network parameters, obtained after model training is shown in Table 4.2. The closeness of prediction between the experimental and ANN model fitted outputs were evaluated by computing the determination coefficient values, as shown by Equation 4.3:

$$R^2 = \left[\frac{\sum_{i=1}^{N} (Y_{modeli} - \overline{Y_{model}})(Y_{observedi} - \overline{Y_{observed}})}{(N-1)\, S_{Y_{model}}\, S_{Y_{observed}}} \right]^2 \qquad (4.3)$$

where, R^2 is the determination coefficient, Y_{modeli} is the model fitted values, $Y_{observedi}$ is the experimentally observed values, N is the number of cases analyzed, $\overline{Y}$ is the average value, and S_Y is the standard deviation.

Absolute average sensitivity was calculated by summing the changes in the output variables caused by moving the input variables by a small amount over the entire training set. This was estimated according to Equation 4.4:

$$S_{ki,\,abs} = \frac{\sum_{p=1}^{p} \left| S_{ki}^{(p)} \right|}{p} \qquad (4.4)$$

where, $S_{ki}^{(p)}$ is the sensitivity of a trained output, and p is the number of training patterns.

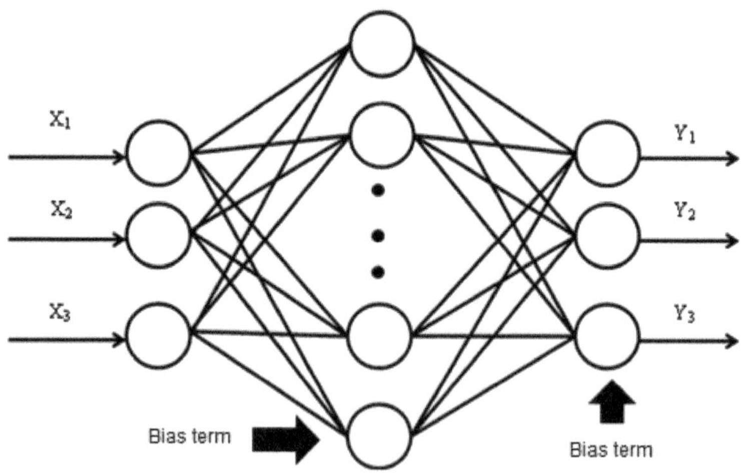

Figure 4.2 Schematic of the three-layered artificial neural network

Table 4.2 Range of neural network parameters used during training of the ANN

Parameter	Range of values tested	Best values obtained
Hidden Neurons (N_H)	3 - 8	7
Training cycles (T_c)	20,000 - 70,000	22,357
Learning rates (η)	0.1 - 0.9	0.9
Training number (N_{Tr})		69
Testing number (N_{Te})		30
Momentum (α)	0.1 - 0.9	0.9
Epoch size (ε)	50	
Confidence interval	95%	
Error of tolerance	0.0001	

4.3 Results

4.3.1 Batch experiments

Batch tests were conducted to access the activities and the effects of pH on SR and to compare sulfate reducing efficiencies of granular sludge with attached biofilm from the IFB (Figure 4.3). The overall sulfate reducing activities from the sludge were lower than the attached biomass for the initial pH of 5.0 to 7.0, while there was not much difference observed at pH 3.0 and 4.0. The COD and sulfate removal efficiencies of the attached biomass at pH 7.0 were 47.9 and 56.7 %, which was in a similar range for pH 6.0 (47.7 and 55.3%) and 5.0 (32.5 and 50.0%). For pH 3.0 and 4.0, the COD removal efficiency did not exceed 20.0%, while the sulfate removal efficiency was below 15.0%. The final pH values after 28 days of incubation are shown in Table 4.3.

Table 4.3 Average pH after 28 days of incubation in the batch experiments using inoculum (granular sludge) and IFB biomass

	Electron acceptor	pH 7.0	pH 6.0	pH 5.0	pH 4.0	pH 3.0
Granular sludge	Sulfate	6.48	6.24	5.35	4.4	3.43
IFB PE bead	Sulfate	6.42	6.44	5.34	4.52	3.77
	Thiosulfate	6.42	6.44	5.34	4.52	3.77

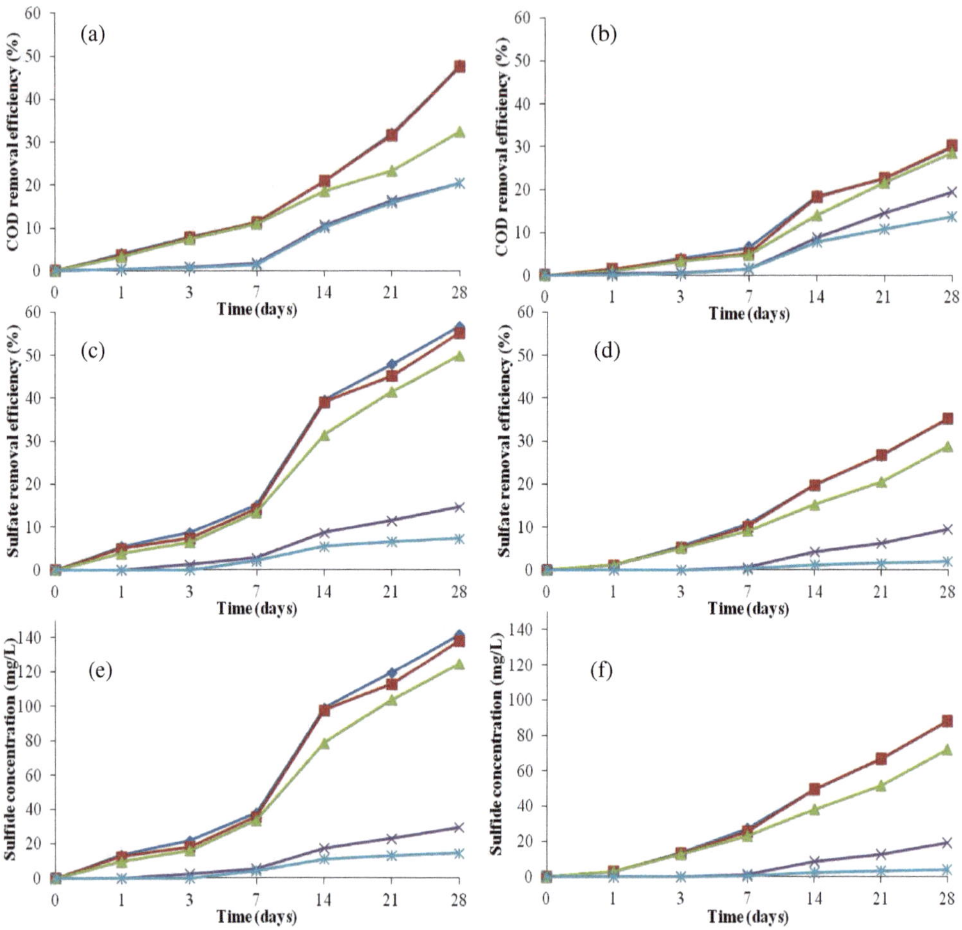

Figure 4.3 COD and sulfate removal efficiencies as well as sulfide production of attached biomass on the IFB PE bead (a, c, e), and the granular sludge inoculum (b, d, f).

(♦) pH 7.0, (■) pH 6.0, (▲) pH 5.0, (×) pH 4.0 and (✱) pH 3.0

4.3.2 Reactor performance

4.3.2.1 COD, sulfate and thiosulfate removal

The reactor operation using sulfate as electron acceptor at pH 7.0 gave average removal efficiencies of COD and sulfate of 75.0% and 74.4%, respectively. The removal efficiencies dramatically dropped when decreasing the pH to 5.0 in R2. The efficiency at pH 7.0 was quite stable throughout the study as shown in Figures 4.4a and 4.4b. However, at pH 5.0 of the period Sulfate I, the efficiencies dropped to 18.3% and 23.3% for COD and sulfate removal, respectively (Figure 4.4b). As a consequence, the reactor that was operated at low pH was stopped and reverted back to pH 7.0 until the sulfate reducing activity was quite stable again, which was 29 days (Recovery period) before the next period began (Sulfate II period). The result obtained in the Sulfate II period, which was after the recovery period from the low pH of Sulfate I, showed significant differences at pH 5.0 for both COD and sulfate removal. The average COD and sulfate removal efficiencies at pH 7.0 decreased to 64.0 and 65.8%, while the COD and sulfate removal efficiencies were 58.0 and 50.4% at pH 5.0 in the Sulfate II period, respectively.

For the Thiosulfate period, the sulfate concentration was averagely ~123 mg/L in the freshly prepared influent and ~116 mg/L at the entry point of the reactor column during the operation at pH 7.0. The thiosulfate removal efficiencies were 75.3, 66.3, 62.0%, while the COD removal efficiencies were 63.5, 57.6, 54.8% at pH 7.0, 6.0 and 5.0, respectively (Figure 4.4). Moreover, as the pH was decreased to pH 5.0, a higher sulfate concentration was observed in the reactor effluent.

4.3.2.2 Sulfide production

The average sulfide concentration using sulfate as electron acceptor in R2 was 246.3, 150.4, and 73.7 mg/L and 202.7, 174.2 and 150.7 mg/L in periods Sulfate I and II at pH 7.0, 6.0 and 5.0, respectively. The optimal sulfide production at pH 7.0 was 246.3 mg/L throughout the experimental period of R1. When thiosulfate was used as the electron acceptor, the average sulfide production in R2 was 301.3, 214.6, 162.7 mg/L at pH 7.0, 6.0 and 5.0, respectively.

4.3.2.3 Acetate toxicity tests

High acetate concentrations can be toxic to the SRB at low pH, thus this study was performed by adding 100.0 to 500.0 mg/L acetate to the ethanol proportionally in order to make up for 1 g/L COD. Figure 5 shows that at pH 7.0 (Figure 4.5c, 4.5d), the COD and sulfate removal efficiencies were higher than at pH 5.0 (Figure 4.5a, 4.5b). At day 14, the maximum removal efficiencies at pH 7.0 with an acetate concentration of 100 mg/L were ~30% and 28% for COD and sulfate, respectively. However, at higher acetate concentrations, the SR efficiencies were decreased accordingly. The sulfate removal efficiency dropped to approximately 8%, while the COD removal efficiency was 10% at pH 7.0 in the treatment with 400 and 500 mg/L acetate.

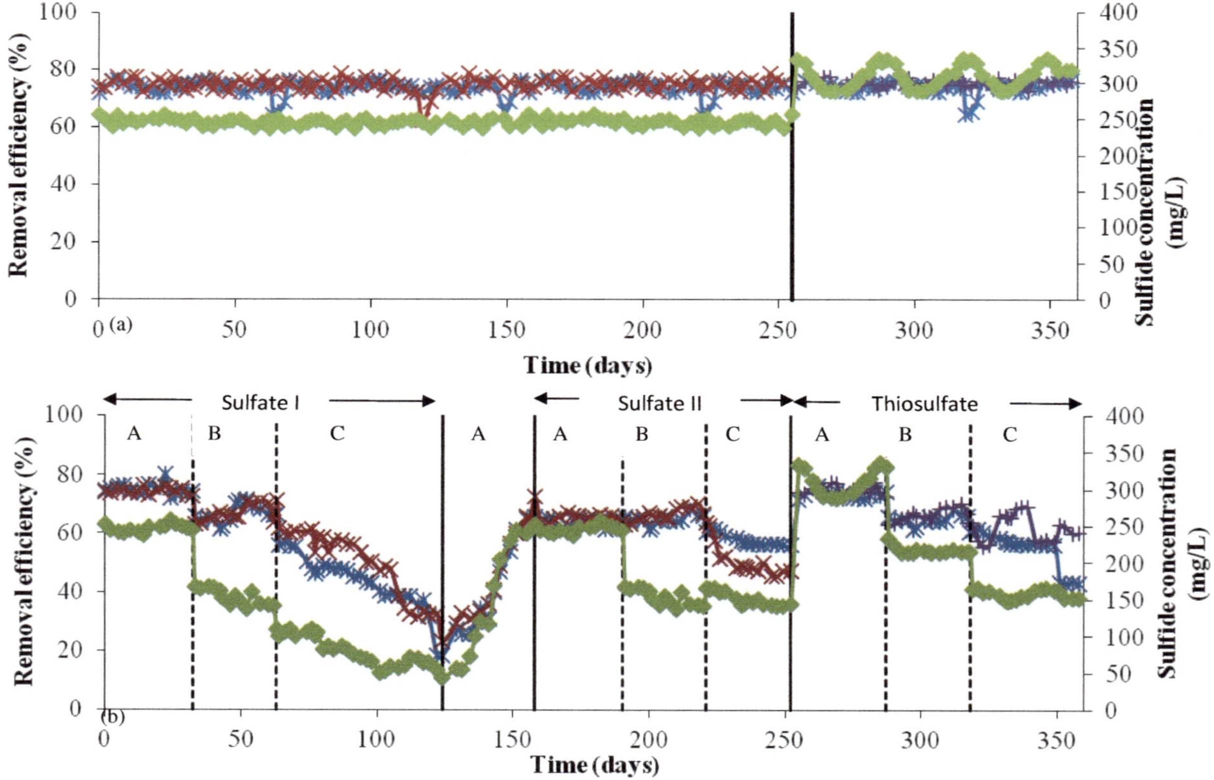

Figure 4.4 SR activities in a) R1 and b) R2 while A = pH 7.0, B = pH 6.0 and C = pH 5.0. (×) COD, (×) Sulfate, (♦) Sulfide and (|) Thiosulfate

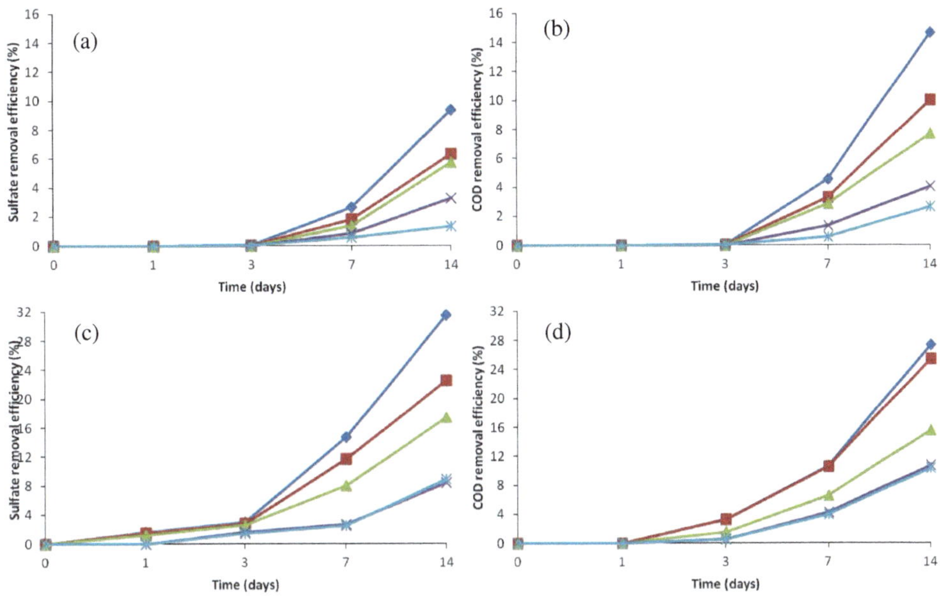

Figure 4.5 Sulfate (left) and COD (right) removal efficiency with different acetate concentrations at pH 5.0 (a, b) and at pH 7.0 (c, d). (♦) 100, (■) 200, (▲) 300, (×) 400 and (⋇) 500 mg/L

4.3.2.4 Artificial neural network modeling

In order to evaluate the relationship between input (operated pH, Influent COD and sulfate) and output (sulfide production, COD and sulfate removal) parameters, the data obtained from the Sulfate I and II phases were modeled using an ANN approach. After trial and error approximation of the internal network parameters, the best network topology was ascertained as 3-7-3, which corresponds to the number of neurons in the input layer, hidden layer and output layer, respectively. The experimental and ANN fitted COD and sulfate removal efficiencies and sulfide produced in the IFB for the training data is shown in Figures 4.6a, 4.6b and 4.6c, respectively. It is apparent that, while predicting these profiles, the network was able to exactly map the data points. However, a few data points were not adequately mapped by the network during training. This was presumable due to a severe drop in these values when the IFB was operated at pH 5.0. The COD and sulfate removal efficiencies dropped to values <50%, while the sulfide produced also dropped to <100 mg/L. This decrease in COD and sulfate removal and sulfide production profiles would have caused an impact in the networks generalization pattern while predicting the performance parameters, a pattern that has often been reported during ANN modeling of bioprocesses. However, the neural network model showed good predictive ability for performance variables with R^2 values of 0.86, 0.86 and 0.93, respectively for COD removal, sulfate removal and sulfide production.

After training, the network was provided with a separate set of data consisting of 30 data points for testing the developed model. These results are illustrated in Figures 4.6d, 4.6e and 4.6f, respectively. A comparison between the experimental and ANN model fitted profiles for COD removal, sulfate removal and sulfide produced reveals the predictive capability of the model. The model was able to adequately identify the low and high peaks of these profiles with R2 values of 0.78, 0.81 and 0.91, respectively, for COD removal, sulfate removal and sulfide production. Table 4.4 shows the different connection weights, *i.e.*, connection weights between input to hidden layer, and hidden to output layer for the developed 3-7-3 network topology together with the bias term.

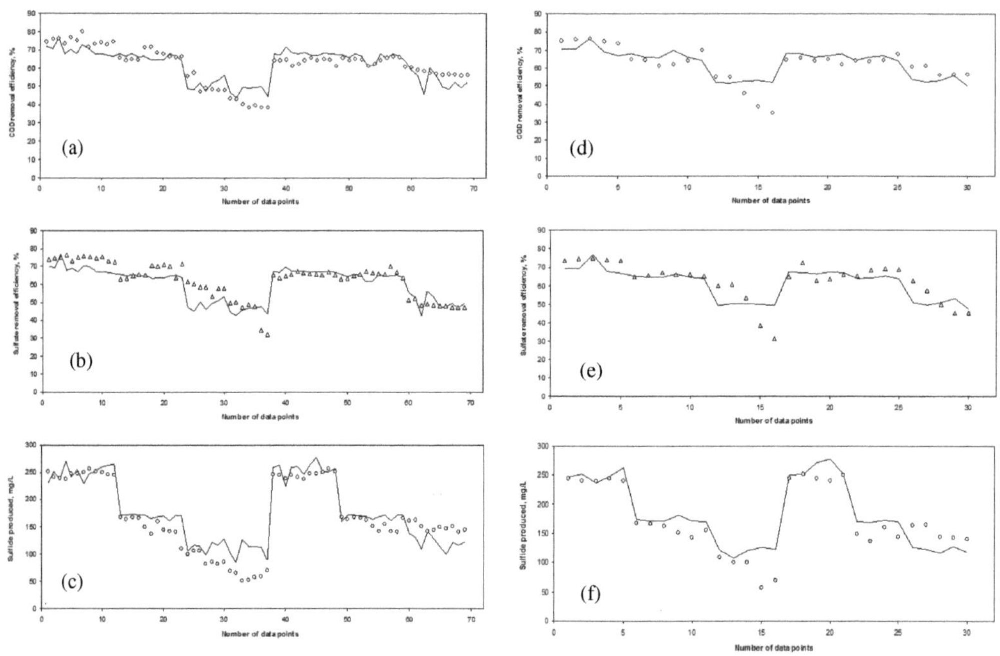

Figure 4.6 Experimental and ANN fitted profiles of COD (a, d) and sulfate (b, e) removal efficiency and sulfide production (c, f) in the training and test data sets. (Δ) Experimental and (——) ANN fitted

Two-dimensional contour plots, provided by the software NNMODEL, helped to identify the influence of operational parameters on the model outputs. These plots can also provide adequate insight in the different process conditions required to obtain the best removal efficiencies for COD and sulfate and the conditions that will maintain IFB operation within non-inhibitory sulfide levels. Figures 4.7a, 4.7b and 4.7c show the sulfide production as well as COD and sulfate removal efficiency profiles as a function of influent COD and sulfate concentration. These profiles were generated at a constant pH value of 7.0. These contours can be interpreted as follows: (i) high concentrations of sulfide (232 mg/L) are produced in the IFB when sulfate concentrations exceed 700 mg/L and COD concentrations exceed 650 mg/L, (ii) COD removal efficiencies can be maintained at > 83%, when the influent COD concentration is < 800 mg/L and sulfate concentration is > 600 mg/L, and (iii) sulfate removal efficiencies > 83% can be achieved when the COD concentration is < 800 mg/L, irrespective of the influent sulfate concentrations.

The results of the sensitivity analysis, measured as absolute average sensitivity values (Figure 4.8), showed that the pH was the most important factor affecting all output parameters (COD removal, sulfate removal and sulfide production). The influent sulfate concentration was ranked as the second impact factor, while COD was ranked third.

4.4 Discussion

4.4.1 Sulfate reducing activities at pH 5.0

In this study, the COD and sulfate removal efficiencies at pH 7.0 were quite stable at 75.0 and 74.4% for reactor R1 (Figure 4.4a). However, decreasing the pH in the reactor drastically affected the efficiency, especially at pH 5.0, as can be seen in the period Sulfate I (Figure 4.4b). This can be due to the activity of the inoculum used in this study, which was not subjected to low pH conditions before it was inoculated into the reactor. Thus, when the pH in the R1 reaches lower values, the activity of SR was suppressed by the low pH or the SR were not able to survive. Another possible explanation for the low bioreactor performance could be due to the formation of toxic compounds such as sulfide. Sulfide has been reported in many cases to be toxic to SRB at elevated concentrations, especially in non-dissociation form (H_2S) (Al Zuhair et al., 2008; O'Flaherty et al., 1999). Activity at a lower pH has nevertheless been reported in the literature, despite the lower efficiencies at below pH 5.0 (Costa et al., 2008; Martins et al., 2009; Neculita et al., 2007). Moreover, apart from the

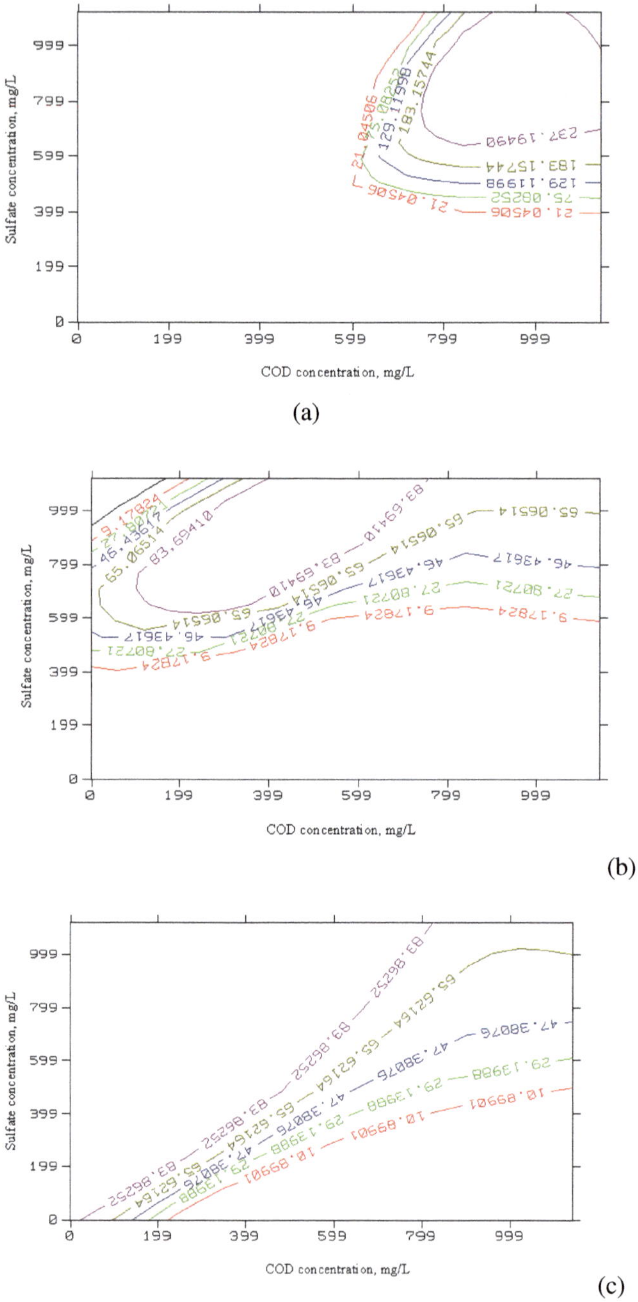

Figure 4.7 Contour plots showing the effect of different input parameters on the sulfide production (a), COD removal (b) and sulfate removal (c). The pH was kept at constant at 7.0 while obtaining these contours.

Table 4.4 Connection weights of the developed neural network model (3-7-3)

Input layer to hidden layer

	HID1	HID2	HID3	HID4	HID5	HID6	HID7
X_1	-10.261	-0.017	-6.150	-19.863	-22.884	-5.678	-5.485
X_2	14.559	-8.527	-2.077	-12.277	11.198	-5.655	-4.390
X_3	-21.183	-3.612	-4.947	32.728	3.759	6.660	0.260
Bias	7.506	5.557	-0.632	-3.084	1.052	-0.364	-0.286

Hidden layer to output layer

	Y_1	Y_2	Y_3	
				HID1 to HID7: Hidden layer neurons
HID1	-11.532	-11.611	-7.068	Bias: Bias term
HID2	-9.244	-9.687	-12.961	**Input to model**
HID3	-2.934	0.496	2.656	X_1: Influent COD concentration, mg/L
HID4	-0.202	-0.481	2.397	X_2: Influent SO_4^{2-} concentration, mg/L
HID5	5.181	4.904	-10.708	X_3: Operating pH
HID6	-4.907	4.236	-1.112	**Output of the model**
HID7	-1.102	1.483	2.995	Y_1: COD removal efficiency, %
Bias	1.335	1.240	0.909	Y_2: SO_4^{2-} removal efficiency, %
				Y_3: Total sulfide production, mg/L

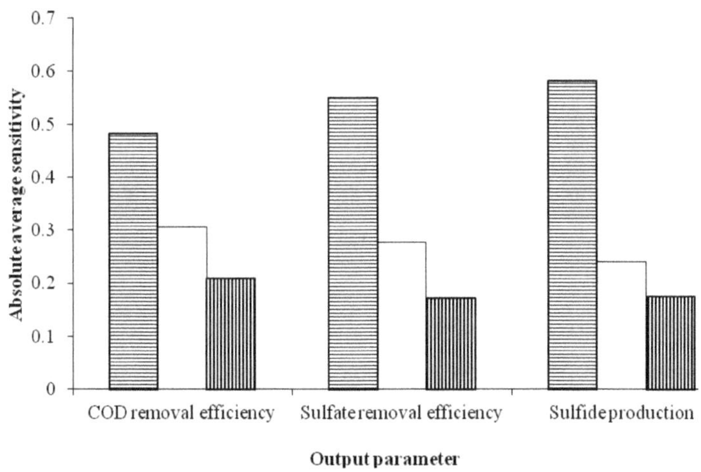

Figure 4.8 Sensitivity analysis of input parameters represented by absolute average sensitivity values. (⊟) pH, (□) Sulfate and (▥) COD

pH dependence of SR activities, sulfide speciation is a pH based phenomenon. S^{2-} and HS^- are formed at the basic and neutral pH range, while H_2S, which is toxic to SRB, is the dominant species at acidic pH (Cohen, 2006; Jong et al., 2006; Sawyer et al., 2003).

Apart from the microbiological factors, also the electron donor contributes to the efficiency of the reactor. According to Figure 4.3, the higher the acetate concentration in the system, the lower is the COD and sulfate removal efficiency. Acetate concentrations > 300 mg/L decreased the efficiency and activity of the biofilm. These results are in close agreement to the observations made during the Sulfate I period at pH 5.0 (Figure 4.4b) in reactor R2. The remaining COD from ethanol degradation mainly comes from acetate which many SRB cannot fully oxidize to CO_2 (Equation 4.5). Thus, its accumulation in the system can cause toxicity (Kaksonen et al., 2004; Lens et al., 2002):

$$2\ C_2H_5OH + SO_4^{2-} \rightarrow 2\ CH_3COO^- + HS^- + H^+ + 2\ H_2O \qquad (4.5)$$

The lowest pH at which sulfate reducing activities still occurred in this study was pH 5.0 (±0.2) using ethanol as the electron donor. The high acetate load can alter the growth and activities of bacterial communities. Several published reports have shown that lactate as electron donor can make a SRB bioreactor fully functional at even lower pH (pH 4.0) (Elliott et al., 1998; Villa-Gomez et al., 2011). However, ethanol as electron donor supports high sulfate reduction efficiencies (> 80%), is relatively cheap and easy to use compared to lactate (Liamleam et al., 2007a; Smul et al., 1997). According to Figure 4.4b, once the operating conditions were restored to their original values, the strains which were able to tolerate or adapt to a low pH environment became the dominant species and supported higher SR activities at pH 5.0 during the Sulfate II period compared to the activities in Sulfate I period at the same pH.

4.4.2 Thiosulfate as electron acceptor

During thiosulfate disproportionation, according to Equation 6, other sulfur compounds such as sulfide and sulfate are also generated (Bak et al., 1987). The latter can be reduced to sulfide by SRB.

$$S_2O_3^{2-} + H_2O \rightarrow SO_4^{2-} + HS^- + H^+ \qquad (4.6)$$

This process does not require an electron donor or acceptor (Jørgensen, 1990; Jørgensen et al., 1991). At low pH, thiosulfate can disproportionate and yield other sulfur compounds in any of the three possible mechanisms as shown in Equations 7, 8 and 9 (Mizoguchi et al., 1976):

$$5\ S_2O_3^{2-} + 6\ H^+ \rightarrow 2\ S + 2\ S_4O_6^{2-} + 3\ H_2O \qquad (4.7)$$

$$S_2O_3^{2-} + H^+ \rightarrow S + HSO_3^- \qquad (4.8)$$

$$3\ S_2O_3^{2-} + 2\ H^+ \rightarrow 4S + 2\ SO_4^{2-} + H_2O \qquad (4.9)$$

These equations clearly explain the trend observed in Figure 4a, wherein the production of total sulfide from thiosulfate was higher than from sulfate. Thiosulfate disproportionation to sulfate was also observed inside the influent tank (122.7 mg/L), excluding the 53.1 mg/L sulfate in tap water which was used to prepare the medium. In addition, at the point where the influent entered the column, the sulfate concentration was 115.8 mg/L. This change in sulfate concentration might be caused by enzymatic activities, for instance *Salmonella typhimurium* can produce H_2S from thiosulfate or sulfite using thiosulfate reductase or sulfite reductase, respectively (Alami et al., 1995; Barrett et al., 1987; Hallenbeck et al., 1989).

4.4.3 IFB bioreactor technology application

For applications in wastewater treatment, one of the common uses of SRB bioreactors is the removal and recovery of heavy metals from wastewater (Bijmans et al., 2011). There are many studies that have shown improvement in the efficiency of SRB reactors which resulted in new bioreactor configurations that are applied nowadays at full scale (Kaksonen et al., 2007). Table 4.5 presents an overview of the different bioreactor configurations that have been used to treat wastewater using different electron donors. The IFB reactor is one of the developed configurations which have proven to be promising over others as a single unit (Villa-Gomez et al., 2011). The key of this reactor is a biofilm on a floatable carrier material in the down-flow circulation which allows the metal precipitates to be collected at the bottom of the reactor by gravitation (Celis et al., 2009; Villa-Gomez et al., 2011). Although, the IFB has already been applied for treatment of highly loaded organic wastewater, there are still many potential applications for metal contaminated wastewater treatment. It has been reported that an IFB has the potential to be used for high organic rich wastewater like distillery effluent, achieving ~84% COD removal at an organic loading rate (OLR) of 35 kg COD/m^3/d (Sowmeyan et al., 2008a; Sowmeyan et al., 2008b). In recent studies focusing on the use of IFB for metals (Cd, Cu, Pb and Zn) removal, it was shown that the IFB can achieve >90% metal removal efficiency (Villa-Gomez et al., 2011; Villa-Gomez et al., 2015). In those studies, the sulfide concentration and HRT were found to be important factors determining the location and size of the precipitate.

Table 4.5 COD and sulfate removal efficiencies of several bioreactor configurations

e^- donor	Reactor type	Removal efficiency (%) COD	SO_4^{2-}	$S_2O_3^{2-}$	Sulfide production (mg/L)	References
Ethanol	IFB (pH 5.0)	58.0	50.4	na	150.7	This study
	IFB (pH 7.0)	75.0	74.4	na	246.3	
Ethanol	IFB (pH 5.0)	58.01	na	62.21	161.34	
	IFB (pH 7.0)	73.46	na	74.91	246.92	
Acetate	SMBR	~100	na	95.0	na	Pikaar et al. (2013)
Lactate	IFB	15	56	na	na	Villa-Gomez et al. (2011)
Glucose	IFB	84	na	na	na	Sowmeyan and Swaminathan (2008a)
Starch wastewater	FBR	72.8	na	na	na	Rajasimman and Karthikeyan (2007)
VFA	DFFB	70 - 80	79	na	207	Celis-García et al. (2007)

Note: na- not available

There was a whitish yellow precipitate present in the effluent tubes, indicating that the dissolved sulfide was converted into elemental sulfur (S^0). Indeed, the excess sulfide produced in the bioreactor when exposed to the air can be converted biologically or chemically into sulfate or elemental sulfur *via* sulfide oxidation process (Lens et al., 2002; Sahinkaya et al., 2011; Zhang et al., 2008). Metal recovery can be coupled to S^0 recovery, which can be used in other industrial processes, *i.e.*, sulfuric acid production or as an electron donor for denitrification (Dutta et al., 2008; Liamleam et al., 2007b; Wang et al., 2011).

4.5 Conclusions

The use of ethanol as an electron donor allowed IFB reactor operation at a pH as low as 5.0 using sulfate as the electron acceptor. The remaining acetate during the operation at pH 5.0 affects the performance of sulfate reducing activity. The results from ANN modeling showed that a three-layered back propagation neural network (3-7-3) has a great capacity to establish the relationship between the performance indices, namely COD removal, sulfate removal and sulfide production, and IFB operational parameters, namely pH, influent COD and sulfate concentrations. Moreover, the sensitivity analysis showed that the pH was the most important parameter affecting the longevity and performance of the sulfate reducing IFB using both sulfate and thiosulfate as the electron acceptor.

References

Al Zuhair, S., H El-Naas, M. and Al Hassani, H., 2008. Sulfate inhibition effect on sulfate reducing bacteria. J. Biochem. Technol., 1(2): 39-44.

Alami, N. and Hallenbeck, P.C., 1995. Cloning and characterization of a gene cluster, phsBCDEF, necessary for the production of hydrogen sulfide from thiosulfate by *Salmonella typhimurium.* Gene, 156(1): 53-57.

APHA, 1999. *Standard Methods for the Examination of Water and Wastewater. -- 20th Edition/ Prepared and Published Jointly by American Public Health Association, American Water Works Association, Water Environment Federation.* AWWA.

Atasoy, A.D., Babar, B. and Sahinkaya, E., 2013. Artificial neural network prediction of the performance of upflow and downflow fluidized bed reactors treating acidic mine drainage water. Mine Water Environ., 32(3): 222-228.

Bak, F. and Cypionka, H., 1987. A novel type of energy metabolism involving fermentation of inorganic sulphur compounds. Nature, 326(6116): 891-892.

Barrett, E.L. and Clark, M.A., 1987. Tetrathionate reduction and production of hydrogen sulfide from thiosulfate. Microbiol. Rev., 51(2): 192-205.

Bayrakdar, A., Sahinkaya, E., Gungor, M., Uyanik, S. and Atasoy, A.D., 2009. Performance of sulfidogenic anaerobic baffled reactor (ABR) treating acidic and zinc-containing wastewater. Bioresour. Technol., 100(19): 4354-4360.

Bijmans, M.F.M., Buisman, C.J.N., Meulepas, R.J.W. and Lens, P.N.L., 2011. Sulfate reduction for inorganic waste and process water treatment. Compr. Biotechnol.: 435-446.

Cao, J., Zhang, G., Mao, Z.-S., Li, Y., Fang, Z. and Yang, C., 2012. Influence of electron donors on the growth and activity of sulfate-reducing bacteria. Int. J. Miner. Process., 106-109(0): 58-64.

Celis-García, L.B., Razo-Flores, E. and Monroy, O., 2007. Performance of a down-flow fluidized bed reactor under sulfate reduction conditions using volatile fatty acids as electron donors. Biotechnol. Bioeng., 97(4): 771-779.

Celis, L., Villa-Gómez, D., Alpuche-Solís, A., Ortega-Morales, B. and Razo-Flores, E., 2009. Characterization of sulfate-reducing bacteria dominated surface communities during start-up of a down-flow fluidized bed reactor. J. Ind. Microbiol. Biotechnol., 36(1): 111-121.

Cohen, R.R.H., 2006. Use of microbes for cost reduction of metal removal from metals and mining industry waste streams. J. Clean. Prod., 14(12-13): 1146-1157.

Cord-Ruwisch, R., 1985. A quick method for the determination of dissolved and precipitated sulfides in cultures of sulfate-reducing bacteria. J. Microbiol. Methods, 4(1): 33-36.

Costa, M., Martins, M., Jesus, C. and Duarte, J., 2008. Treatment of acid mine drainage by sulphate-reducing bacteria using low cost matrices. Water Air Soil Pollut., 189(1): 149-162.

Dutta, P.K., Rabaey, K., Yuan, Z. and Keller, J., 2008. Spontaneous electrochemical removal of aqueous sulfide. Water Res., 42(20): 4965-4975.

Eaton, A.D., Franson, M.A.H., Association, A.P.H., Association, A.W.W. and Federation, W.E., 2005. *Standard Methods for the Examination of Water & Wastewater*. American Public Health Association, Washington D.C., the United States.

Elliott, P., Ragusa, S. and Catcheside, D., 1998. Growth of sulfate-reducing bacteria under acidic conditions in an upflow anaerobic bioreactor as a treatment system for acid mine drainage. Water Res., 32(12): 3724-3730.

González - Sánchez, A., Meulepas, R. and Revah, S., 2008. Sulfur formation and recovery in a thiosulfate - oxidizing bioreactor. Environ. Technol., 29(8): 847-853.

Hallenbeck, P.C., Clark, M.A. and Barrett, E.L., 1989. Characterization of anaerobic sulfite reduction by *Salmonella typhimurium* and purification of the anaerobically induced sulfite reductase. J. Bacteriol., 171(6): 3008-15.

Heidelberg, J.F., Seshadri, R., Haveman, S.A., Hemme, C.L., Paulsen, I.T., Kolonay, J.F., Eisen, J.A., Ward, N., Methe, B., Brinkac, L.M., Daugherty, S.C., Deboy, R.T., Dodson, R.J., Durkin, A.S., Madupu, R., Nelson, W.C., Sullivan, S.A., Fouts, D., Haft, D.H., Selengut, J., Peterson, J.D., Davidsen, T.M., Zafar, N., Zhou, L., Radune, D., Dimitrov, G., Hance, M., Tran, K., Khouri, H., Gill, J., Utterback, T.R., Feldblyum, T.V., Wall, J.D., Voordouw, G. and Fraser, C.M., 2004. The genome sequence of the anaerobic, sulfate-reducing bacterium *Desulfovibrio vulgaris* Hildenborough. Nat Biotech, 22(5): 554-559.

Jong, T. and Parry, D.L., 2003. Removal of sulfate and heavy metals by sulfate reducing bacteria in short-term bench scale upflow anaerobic packed bed reactor runs. Water Res., 37(14): 3379-3389.

Jong, T. and Parry, D.L., 2006. Microbial sulfate reduction under sequentially acidic conditions in an upflow anaerobic packed bed bioreactor. Water Res., 40(13): 2561-2571.

Jørgensen, B.B., 1990. A thiosulfate shunt in the sulfur cycle of marine sediments. Science, 249(4965): 152-154.

Jørgensen, B.B. and Bak, F., 1991. Pathways and microbiology of thiosulfate transformations and sulfate reduction in a marine sediment (Kattegat, Denmark). Appl. Environ. Microbiol., 57(3): 847-856.

Kaksonen, A.H., Plumb, J.J., Franzmann, P.D. and Puhakka, J.A., 2004. Simple organic electron donors support diverse sulfate-reducing communities in fluidized-bed reactors treating acidic metal- and sulfate-containing wastewater. FEMS Microbiol. Ecol., 47(3): 279-289.

Kaksonen, A.H. and Puhakka, J.A., 2007. Sulfate reduction based bioprocesses for the treatment of acid mine drainage and the recovery of metals. Eng. Life Sci., 7(6): 541-564.

Lens, P. and Kuenen, J., 2001. The biological sulfur cycle: novel opportunities for environmental biotechnology. Water Sci. Technol., 44(8): 57-66.

Lens, P.N.L. and Pol, L.H., 2000. *Environmental Technologies to Treat Sulfur Pollution: Principles and Engineering*. IWA Publishing, London, United Kingdom.

Lens, P., Vallerol, M., Esposito, G. and Zandvoort, M., 2002. Perspectives of sulfate reducing bioreactors in environmental biotechnology. Rev. Environ. Sci. Biotechnol., 1(4): 311-325.

Li, S., Wang, Z. and Fei, S., 2009. Finite-time control of a bioreactor system using terminal sliding mode. Int. J. Innov. Comput. I., 5(10B): 3495-3504.

Liamleam, W. and Annachhatre, A.P., 2007a. Electron donors for biological sulfate reduction. Biotechnol. Adv., 25(5): 452-463.

Liamleam, W. and Annachhatre, A.P., 2007b. Treating industrial discharges by thermophilic sulfate reduction process with molasses as electron donor. Environ. Technol., 28(6): 639-647.

Liang, F., Xiao, Y. and Zhao, F., 2013. Effect of pH on sulfate removal from wastewater using a bioelectrochemical system. Chem. Eng. J., 218(0): 147-153.

Liljeqvist, M., Sundkvist, J.-E., Saleh, A. and Dopson, M., 2011. Low temperature removal of inorganic sulfur compounds from mining process waters. Biotechnol. Bioeng., 108(6): 1251-1259.

Martins, M., Faleiro, M.L., Barros, R.J., Veríssimo, A.R., Barreiros, M.A. and Costa, M.C., 2009. Characterization and activity studies of highly heavy metal resistant sulphate-reducing bacteria to be used in acid mine drainage decontamination. J. Hazard. Mater., 166(2-3): 706-713.

Mizoguchi, T., Takei, Y. and Okabe, T., 1976. Chemical behavior of low valence sulfur compounds. X. Disproportionation of thiosulfate, trithionate, tetrathionate and sulfite under acidic conditions. Bull. Chem. Soc. Jpn, 49(1): 70-75.

Mjalli, F.S. and Al-Asheh, S., 2005. Neural-networks-based feedback linearization versus model predictive control of continuous alcoholic fermentation process. Chem. Eng. Technol., 28(10): 1191-1200.

Mullai, P., Arulselvi, S., Ngo, H.-H. and Sabarathinam, P.L., 2011. Experiments and ANFIS modelling for the biodegradation of penicillin-G wastewater using anaerobic hybrid reactor. Bioresour. Technol., 102(9): 5492-5497.

Nagpal, S., Chuichulcherm, S., Peeva, L. and Livingston, A., 2000. Microbial sulfate reduction in a liquid-solid fluidized bed reactor. Biotechnol. Bioeng., 70(4): 370-380.

Neculita, C.-M., Zagury, G.J. and Bussière, B., 2007. Passive treatment of acid mine drainage in bioreactors using sulfate-reducing bacteria. J. Environ. Qual., 36(1): 1-16.

O'Flaherty, V. and Colleran, E., 1999. Effect of sulphate addition on volatile fatty acid and ethanol degradation in an anaerobic hybrid reactor. I: process disturbance and remediation. Bioresour. Technol., 68(2): 101-107.

Ongen, A., Kurtulus Ozcan, H. and Arayıcı, S., 2013. An evaluation of tannery industry wastewater treatment sludge gasification by artificial neural network modeling. J. Hazard. Mater., 263, Part 2(0): 361-366.

Pikaar, I., Meulepas, R.J., Jagersma, C.G., Stams, A.J. and Lens, P.N., 2013. Thiosulfate removal by disproportionation in anaerobic thermophilic bioreactors, G16 Conference, pp. 83.

Rajasimman, M. and Karthikeyan, C., 2007. Aerobic digestion of starch wastewater in a fluidized bed bioreactor with low density biomass support. J. Hazard. Mater., 143(1-2): 82-86.

Sahinkaya, E., 2009. Biotreatment of zinc-containing wastewater in a sulfidogenic CSTR: performance and artificial neural network (ANN) modelling studies. J. Hazard. Mater., 164(1): 105-113.

Sahinkaya, E. and Gungor, M., 2010. Comparison of sulfidogenic up-flow and down-flow fluidized-bed reactors for the biotreatment of acidic metal-containing wastewater. Bioresour. Technol., 101(24): 9508-9514.

Sahinkaya, E., Hasar, H., Kaksonen, A.H. and Rittmann, B.E., 2011. Performance of a sulfide-oxidizing, sulfur-producing membrane biofilm reactor treating sulfide-containing bioreactor effluent. Environ. Sci. Technol., 45(9): 4080-4087.

Sawyer, C., McCarty, P. and Parkin, G., 2003. *Chemistry for Environmental Engineering and Science*. McGraw-Hill Education.

Schippers, A. and Sand, W., 1999. Bacterial leaching of metal sulfides proceeds by two indirect mechanisms via thiosulfate or via polysulfides and sulfur. Appl. Environ. Microbiol., 65(1): 319-321.

Smul, A.D., Dries, J., Goethals, L., Grootaerd, H. and Verstraete, W., 1997. High rates of microbial sulphate reduction in a mesophilic ethanol-fed expanded-granular-sludge-blanket reactor. Appl. Microbiol. Biotechnol., 48(3): 297-303.

Sowmeyan, R. and Swaminathan, G., 2008a. Evaluation of inverse anaerobic fluidized bed reactor for treating high strength organic wastewater. Bioresour. Technol., 99(9): 3877-3880.

Sowmeyan, R. and Swaminathan, G., 2008b. Performance of inverse anaerobic fluidized bed reactor for treating high strength organic wastewater during start-up phase. Bioresour. Technol., 99(14): 6280-6284.

Stucki, G., Hanselmann, K.W. and Hürzeler, R.A., 1993. Biological sulfuric acid transformation: Reactor design and process optimization. Biotechnol. Bioeng., 41(3): 303-315.

Villa-Gomez, D., Ababneh, H., Papirio, S., Rousseau, D.P.L. and Lens, P.N.L., 2011. Effect of sulfide concentration on the location of the metal precipitates in inversed fluidized bed reactors. J. Hazard. Mater., 192(1): 200-207.

Villa-Gomez, D.K., Enright, A.M., Lystia, E., Buttice, A., Kramer, H. and Lens, P., 2015. Effect of hydraulic retention time on metal precipitation in sulfate reducing inverse fluidized bed reactors. J. Chem. Technol. Biotechnol., 90(1): 120-129.

Wang, Y., Pavissich, J.P., Sabba, F., Bott, C. and Nerenberg, R., 2011. Elemental sulfur (S°) as a supplemental electron donor for wastewater denitrification. Proc. Water. Environ. F., 2011(15): 1590-1597.

Zhang, L., De Schryver, P., De Gusseme, B., De Muynck, W., Boon, N. and Verstraete, W., 2008. Chemical and biological technologies for hydrogen sulfide emission control in sewer systems: a review. Water Res., 42(1-2): 1-12.

Chapter 5 Inversed Fluidized Bed SRB Bioreactor for Metal Contaminated Wastewater Treatment

The modified version of this chapter was submitted for publication as:
Janyasuthiwong S., Rene E.R., Esposito G. and Lens P.N.L. (2015).Effect of pH on Cu, Ni and Zn removal by biogenic sulfide precipitation in an inversed fluidized bed bioreactor. *Hydrometallurgy, In Press.*

Summary

Mining and metallurgical operations have resulted in the disposal of large quantities of non-treated metal rich effluents into the environment. From a techno-economic and environmental view-point, metal recovery from such waste streams can solve this persisting problem. This study compares a sulfate reducing inverse fluidized bed bioreactor (IFBs) to recover heavy metals (HMs) as metal sulfide from wastewater at neutral and acidic pH. Two IFBs were operated at pH 7.0 and 5.0 to study the effect of the pH on the metal (Cu, Ni and Zn) removal efficiency from a synthetic acid mine drainage containing the metals. The electron donor used in the study was ethanol, supplied at an organic loading rate of 1 g COD/L·d and a COD/sulfate ratio of 1.0. The average sulfide production was 220.9 mg/L and 152.6 mg/L, respectively, at pH 7.0 and pH 5.0 at a hydraulic retention time of 24 h. The Cu and Zn removal efficiencies were > 90% at an initial concentration of 25 mg/L at both operational pH values; however, high concentrations of Ni (25 mg/L) inhibited the sulfate reducing activities. When the influent concentration of each metal was decreased to 10 mg/L, the sulfate reducing efficiency in the IFB was restored. Cu and Zn removal in the IFBs were satisfactorily predicted by the MINTEQ model, while large errors in the Ni removal efficiency were noticed, irrespective of the operational conditions, probably because the MINTEQ model does not take into account NiS polymorphism and binding with the organic compounds present in the biogenic effluent.

5.1 Introduction

The global trend of metal consumption keeps increasing annually, while the global metal resource reservoirs are being depleting as a result of rapid industrialization. Some of the commonly used metal resources, like those of Cu, are expected to be exhausted within a few decades. As a consequence, metal scarcity is becoming a serious issue at the global scale. Wastewater contaminated with metals has become a parallel issue wherein industries such as mining, metal plating, smelting, alloy casting, silver refineries and storage batteries release significant amounts of metal-rich wastewater into water bodies. The presence of these metals in water causes severe adverse effects to the environment and organisms health (Fu and Wang, 2011; Kiran and Thanasekaran, 2011; Wan Ngah and Hanafiah, 2008; Wang and Peng, 2010). There is thus an urgent need to remove these heavy metals (HMs) from the wastewater before ultimate discharge into the environment. Moreover, if this HMs treatment can be coupled to reuse and recovery processes, the metal scarcity problem can be alleviated.

Various chemical and biological techniques have been developed for the removal and recovery of HMs from wastewater. Chemical-based technologies include chemical precipitation, electrolytic recovery, adsorption, ion-exchange and liquid membrane separation. However, these technologies are not economically feasible to be applied at the industrial scale when the effluent contains low concentrations of HMs (Kadirvelu and Goel, 2007). Biobased technologies, on the other hand, such as biological sulfide precipitation have drawn much attraction among researchers for HMs precipitation due to their advantages such as fast reactivity and low residual metal concentration (Lewis, 2010; Peters et al., 1984).

Sulfate reducing bacteria (SRB) are a group of bacteria capable of reducing sulfate into sulfide under anaerobic conditions (Lens and Kuenen, 2001; Liang et al., 2013). These bacteria are the basis of bioreactors treating HM contaminated wastewaters (Fang et al., 2012). There are several reactor configurations presently available to handle metal-rich wastewater streams and conventional reactors consist of two units: the first stage is a sulfide production unit and the second stage a metal precipitation unit. Such reactor configurations double the space requirement and investment costs. An inversed fluidized bed (IFB) reactor is one of the newer proposed reactor configurations which has shown the feasibility for metal removal and recovery from wastewater within a single unit (Gallegos-Garcia et al., 2009; Villa-Gomez et al., 2011). In this reactor configuration, the presence of a light-weight polypropylene packing material acts as the support matrix for the growth of SRB, while the metal-laden wastewater is fluidized downwards. The advantage of this reactor configuration is the easiness of separating the biomass and the metal sulfide precipitates that can be recovered from the bottom of the IFB (Celis et al., 2009; Gallegos-Garcia et al., 2009; Villa-Gomez et al., 2011).

There are several operational parameters of the IFB that need to be optimized in order to achieve stable sulfate reducing (SR) activities and metal precipitation. One of the most important parameters is the pH. SR activity, like any other biological metabolism, is high at a rather wide pH range from more than 5.2 until less than 8.7 with the maximum growth at pH 6.0 - 8.0 (Eccles, 1999; Gutierrez et al., 2009; Neculita et al., 2007). Sulfide speciation (H_2S, HS^- and S^{2-}) within the bioreactor is also pH dependant (Sawyer et al., 2003). For multi-metal precipitation within the bioreactor, each metal has its own optimal pH range wherein some metals may precipitate within a narrow pH range, which may not overlap with that of other metals. From this point of view, sequential precipitation may be possible by manipulating the pH. Thus, the main focus of this study was to evaluate the effect of the pH and potential use of biogenic sulfide for Cu, Ni and Zn removal at an initial concentration of 10 and 25 mg/L in an ethanol fed IFB operated at a hydraulic retention time (HRT) of 24 h. A comparison between the metal removal efficiencies predicted by Visual MINTEQ and the experimental values is also included in this study.

5.2 Material and methods

5.2.1 Medium composition

The synthetic wastewater used in this study had the following chemical composition (in mg/L): $CaCl_2 \cdot 2H_2O$ - 15, KCl - 250, $KHPO_4$ - 200, NH_4Cl - 300, $MgCl_2 \cdot 6H_2O$ - 120 and yeast extract - 20. Ethanol was used as the electron donor, while Na_2SO_4 was used as the electron acceptor at a COD/sulfate ratio of 1.0. The metals (Cu, Ni and Zn) used in this study were supplied as $CuCl_2 \cdot 2H_2O$, $NiCl_2 \cdot 6H_2O$ and $ZnCl_2$. All chemicals used in this study were of analytical grade.

5.2.2 Source of biomass

The biomass used in the IFB reactor was from Biothane Systems International (Delft, the Netherlands) with a volatile suspended solid (VSS) content of 39.8 g/L on wet basis. The carrier material for the attachment of biomass was made of low-density polyethylene beads (Purell Pe 1810E, Basell Polyolifins, the Netherlands) having a mean diameter of 3 mm.

5.2.3 Inversed fluidized bed reactor set up

Two reactors, namely R1 and R2, with a total volume of 2.5 L each (0.08 m in diameter with 1.0 m height) were constructed from a PMMA (polymethyl methacrylate) pipe mounted with removable rubber caps on the top, where provisions for recirculation flow and a pH probe were attached. The influent was circulated via multichannel peristaltic pumps (Watson-Marlow BV, the Netherlands). A magnetic drive pump IWAKI MD-20R-220N (Iwaki Holland BV, the Netherlands) was used to re-circulate the liquid flow within the reactor and this recirculation flow kept the expansion of the carrier material constant, at ~30% of the total reactor volume. A flow equalization tank was connected to each reactor in order to maintain the liquid level in the column.

5.2.4 Metal removal experiment

5.2.4.1 Batch experiment

Batch experiments to determine the HMs removal efficiency using effluent collected from the two reactors were performed in 250 mL Erlenmayer flasks. A known amount of effluent was withdrawn from the reactors R1 and R2 that contained, respectively, 220.9 and 152.6 mg/L sulfide. The initial Cu, Ni and Zn concentrations in the batch experiments were 25 mg/L each. The various metal mixtures and the effluent from R1 and R2 were mixed adequately by placing the flask on a Lab Line Multi-stirrer. Both solutions were mixed with the same volume ratio to make up a total volume of 120 mL. Samples were collected after 1, 5, 10, 30, 60, 90 and 120 min and analyzed for their residual metal concentration. These experiments were done in triplicates and the average values were plotted.

5.2.4.2 Continuous operation

The reactors were operated at room temperature (~25.0 ± 2.0°C). R1 and R2 were fed the same influent at an organic loading rate of 1 g COD/L·d and operated at the same conditions (HRT, biomass expansion, etc) except pH. R1 was operated at pH 7.0 ± 0.2, while R2 was operated at pH 5.0 ± 0.2. The HRT was held constant at 24 h throughout the study. The combinations of metal chloride salts were added in the influent for 14 d, followed by a recovering period of 7 d where no metal was added to the reactors. The recovering period ensured that all the HM components were removed completely as metal precipitates from the liquid phase of the reactor. Table 5.1 shows the combinations and concentrations of individual metals used in the IFB.

Table 5.1 Operational condition of reactor in each period

Conditions	Periods of operation				
	I	II	III	IV	V
Operation time (days)	14	5	14	14	14
Cu (mg/L)	25.0	25.0	10.0	-	10.0
Ni (mg/L)	-	25.0	10.0	10.0	10.0
Zn (mg/L)	25.0	-	-	10.0	10.0

5.2.5 Prediction of metal removal efficiencies

In order to compare the metal removal efficiencies between the theoretical and experimental data obtained from the IFB, Visual MINTEQ Version 3.0 (Gustafsson, 2011) was used to calculate the theoretical metal speciation. The input parameters used to run the model were initial sulfide concentration (average S^{2-} concentration in the recovery period after completion of Periods I, II, III, IV and V, including the period prior to starting the experiment) and individual metal concentrations during each operational period. The output parameters of the MINTEQ model were metal removal efficiencies (based on sulfide speciation) and residual sulfide concentrations.

5.2.6 Analytical methods

Samples (20 ml) from the reactor effluents were collected daily and analyzed for pH, total dissolved sulfide, sulfate, COD and residual metal concentrations. The pH of the reactors were monitored by the program LabVIEW 2009 using sulfide resistant pH electrodes (Prosense, Oosterhout, the Netherlands) attached to a data acquisition card (NI cDAQ-9174, National Instruments, the Netherlands), as described in Villa-Gomez et al. (2014). Total dissolved sulfide was determined using the colorimetric method according to the procedure outlined by Cord-Ruwisch (1985) with a spectrophotometer (PerkinElmer Lambda 20) at 480 nm. COD concentrations were measured by the closed-reflux method (APHA, 1999) and the absorbance was measured with a spectrophotometer at 600.0 nm. The sulfate concentration was measured by ion chromatography (IC) as described by Villa-Gomez et al. (2011). For determining metal concentrations (Cu, Ni and Zn), the samples were filtered, acidified with HNO_3 and measured using an AAnalyst 200 Perkin Elmer Atomic Absorption Spectrometer (AAS) Flame as described by Eaton et al. (2005).

5.3 Results

5.3.1 Batch experiments

Figure 5.1 shows the removal efficiency of Cu, Ni and Zn using the reactor effluents from reactors R1 and R2. The reaction between the metal and sulfide was almost instantaneous wherein the maximum removal efficiencies for the individual HMs was reached within 5 min. The metal removal efficiency using effluent from R1 (pH 7.0) was higher than that with R2 (pH 5.0) with removal efficiencies >95.0 % for all the individual metals (Figure 5.1a), while with R2 effluent, the removal efficiencies were 99.4, 29.5 and 86.4% for Cu, Ni and Zn after 120 min (Figure 5.1b), respectively.

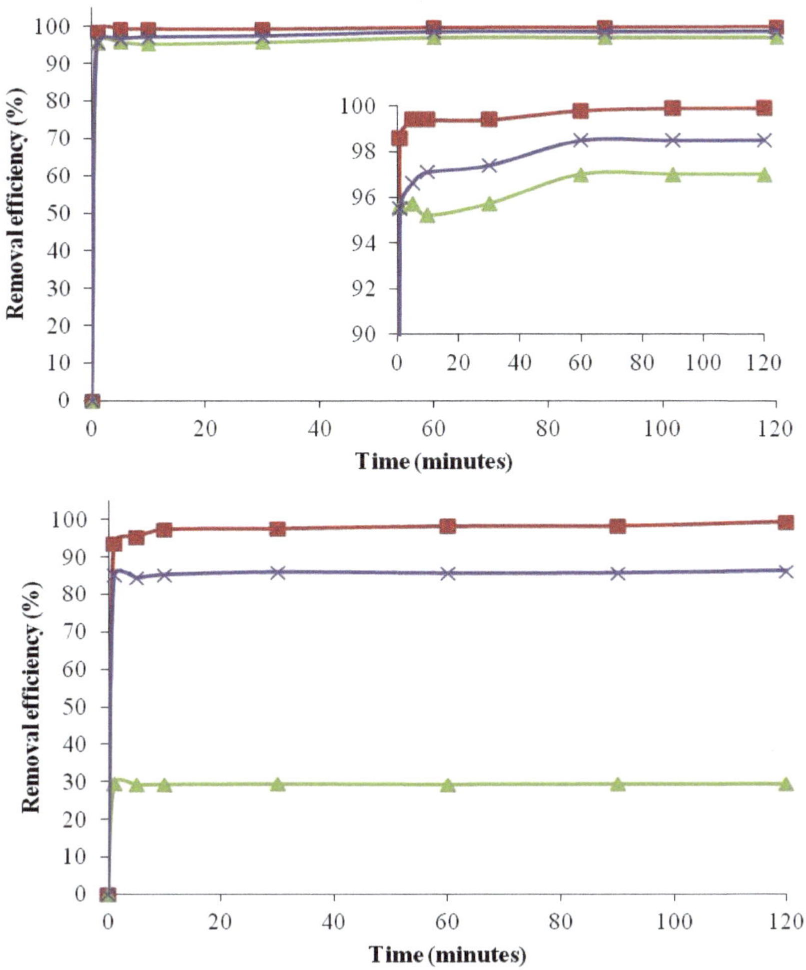

Figure 5.1 Metal removal efficiency with the liquid from IFB reactor at a) pH 7 and b) pH5. (■) Cu, (▲) Ni and (×) Zn

5.3.2 Continuous operation

5.3.2.1 Reactor sulfate reducing performance

The SR activities in reactors R1 and R2 during different periods of operation are shown in Figure 5.2. The average COD and sulfate removal efficiencies (from all recovery periods, including the period prior to Period I) were 72.3 and 73.4% at pH 7.0, while 64.7 and 57.0% at pH 5.0. The average sulfide production was 152.6 and 220.9 mg/L for R1 and R2, respectively.

During Period I, when metals (25 mg/L) were added to the reactors there was a small decline in the reactor performance (COD and SO_4^{2-} removal). In Period II where the Cu and Ni concentration were 25 mg/L each, the reactor operation drastically dropped in both R1 (226.6 to 32.7 mg/L S^{2-}) and R2 (163.5 to 20.4 mg/L S^{2-}). When metal feeding was stopped, the performance of both reactors recovered during a recovery period for 10 d. For Period III, IV and V where 10 mg metals/L were added, the sulfide production ranged between 133.5 and 226.6 mg/L in R1, while between 88.5 and 142.2 mg S^{2-}/L in R2.

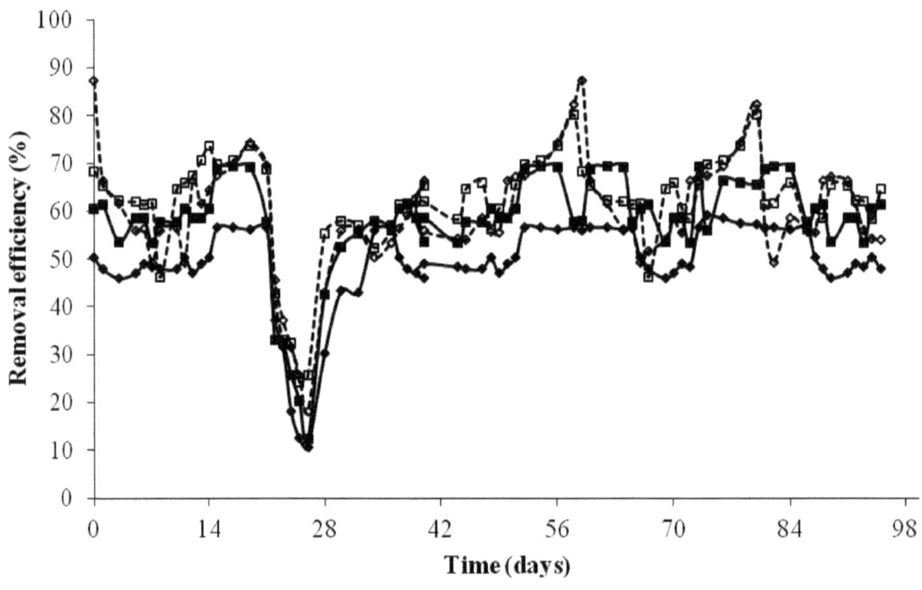

Figure5.2 COD and sulfate removal efficiencies of IFBs. (■) COD pH 5.0 (♦) Sulfate pH 5.0 (□) COD pH 7.0 (◊) Sulfate pH 7.0

5.3.2.2 Metal removal efficiencies

Figure 5.3 shows the HM removal efficiencies during each operational period of reactors R1 and R2. During Period II, with a HM mixture of Cu and Ni at a concentration of 25 mg/L each was fed, the reactor performance dropped drastically to as low as ~5.5%. The IFB operation was stopped on day 26 as a recovery stage of the IFB before continuing to Period III (Cu and Ni at 10 mg/L each). According to Figure 5.3a and 5.3b, Cu removal efficiencies were >95% in both IFB reactors, at both pH 7.0 and 5.0, while Zn with removal efficiencies amounted to >97.0% and ~91.0 - 98.0% at, respectively, pH 7.0 and 5.0. In contrast, the Ni removal efficiency fluctuated over a wider range in both reactors. In Period III, IV and V, the average Ni removal efficiency ranged from 65.9 to 81.5% at pH 7.0 (R1) and from 66.6 to 68.3% at pH 5.0 (R2).

Table 5.2 compares the experimental and MINTEQ predicted removal efficiencies of the individual HMs in reactors R1 and R2. Both the experimental and model predicted values showed no significant differences for Cu and Zn removal. The MINTEQ model predicted that 100% Cu and Zn could be removed as metal sulfides in the reactor and the experimental values were > 90% for both metals. However, large deviations in these values were observed for the Ni removal efficiency, irrespective of the reactor operation (Period II, III, IV and V). In R1, the deviation of the Ni removal efficiency was ~ 20% when Ni was added with another metal, while 34.1% deviation was observed when all three metals were added together. The deviation of the achieved Ni removal efficiency at low pH (R2) when compared to predicted values was quite similar in each period with an average of 32.7%.

5.4 Discussion

5.4.1 IFB reactor performance

This paper shows that biogenic sulfide precipitation in an IFB reactor is promising for treating metal contaminated wastewater. The removal efficiencies of both Cu and Zn exceeded 90%. However, there are some parameters such as pH, organic compounds and metal concentration that affect the performance of SR and HMs removal by the IFB which are discussed below.

The average COD and sulfate removal efficiencies and sulfide production at pH 7.0, i.e., in R1, were higher than those observed at pH 5.0 (R2). This clearly indicates that the pH has a significant effect on the SR activities (Jong and Parry, 2006; Neculita et al., 2007) and sulfide speciation (Sawyer et al., 2003). Moreover, in both IFB reactors, the SR activities dropped once the HMs were added to the IFB influent, especially Ni which had a more toxic effect than Cu and Zn at the same initial concentration. During Period II, with an initial metal

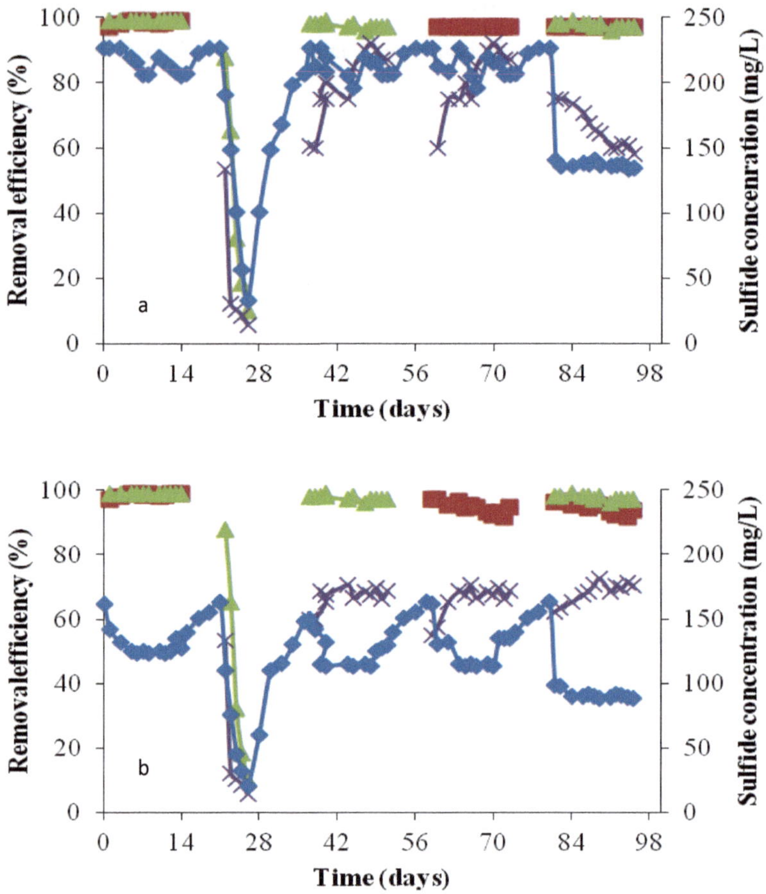

Figure 5.3 Metal removal efficiency in IFB bioreactor a) R1 and b) R2. (♦) Sulfide, (▲) Cu, (×) Ni and (■) Zn

Table 5.2 Comparison of metal removal efficiencies from experiment and model

Period	Initial S^{2-} (mg/L)	Result from prediction model				Average experimental data			
		Cu removal (%)	Ni removal (%)	Zn removal (%)	Remain S^{2-} (mg/L)	Cu removal (%)	Ni removal (%)	Zn removal (%)	Remain S^{2-} (mg/L)
R1									
I	220.7	100	-	100	195.9	99.0	-	98.6	215.7
II	210.4	100	100	-	184.2	45.2	24.22	-	105.8
III	220.5	100	100	-	210.0	97.5	79.7	-	213.1
IV	221.7	-	100	100	211.4	-	81.5	97.0	211.9
V	220.5	100	100	100	205.1	97.5	65.9	97	136.9
R2									
I	150.8	100	-	100	126.0	99.0	-	98.5	130.8
II	148.9	100	100	-	122.7	42.8	18.1	-	57.0
III	152.5	100	100	-	142.0	97.5	66.6	-	124.4
IV	154.4	-	100	100	144.1	-	67.0	97.3	123.1
V	152.5	100	100	100	137.1	97.5	68.3	94.4	91.2

(Cu and Ni) concentration of 25 mg/L each, the metal toxicity (presumably Ni or Ni and Cu interaction) resulted in a lower reactor performance (18.2 and 25.7% for respectively sulfate and COD removal for R1 and 10.6 and 12.2% for respectively sulfate and COD removal for R2). Besides, CuS, NiS and ZnS are formed as complex compounds and tend to build around the biofilm as fine particles where the biogenic sulfide was produced (Villa-Gomez et al., 2011). Thus, Ni may be trapped or could interact with the biofilm surface or penetrate into the biofilm layer and cause toxic effects. This results in much lower metal sulfide formation compared to the MINTEQ model predicted values (100% metal sulfide formation), as the predicted scenario does not take into account toxic effects to SRB. Cabrera et al. (2006) reported that Ni^{2+} concentrations > 8.5 mg/L inhibit *Desulfovibrio* sp. which were incubated for 14 d in batch bottles. However, Ni toxicity towards the SRBs appears reversible, as shown during the recovery period, prior to Period III, wherein the reactor restored its original COD and sulfate removal efficiency and sulfide productivity within 8 - 10 days (day 34 - 36).

5.4.2 Factors influencing metal removal

Although the metal and sulfide reaction is instantaneous, some visible changes were also observed in the two IFBs (*i.e.*, change in liquid color into black). This sulfidic precipitation of HMs is very effective in other configurations, such as an upflow anaerobic sludge blanket (UASB), as well. Gonçalves et al. (2007) reported that SRB in an UASB reactor (HRT of 18 h), using ethanol stillage as electron donor, can reduce Cd and Zn concentrations to below the local discharge standard (0.2 and 5.0 mg/L for Cd and Zn).

Other factors, such as settleability and separation, should also be considered during IFB operation. Parameters such as sulfide concentration and gradient, stoichiometry between the HM and sulfide, shear force in the IFB and the presence of organic compounds also interfere with the settleability of the metal precipitates. According to Table 5.2, the predicted value for the Ni removal efficiency differed from the experimental value possibly due to the sulfide precipitation and environmental factors such as the presence of organic compounds.

5.4.2.1 Metal sulfide formation

The removal efficiencies for Cu and Zn were > 90% in both reactors (pH 7.0 and 5.0). In the pH range of 4.0, the majority of Cu precipitates as CuS (Lewis, 2010). The pH in the operated reactors was higher (7.0 and 5.0, while the effluent pH ranged from 5.9 to 7.3), thus most of the Cu was removed. However, for separation between the liquid and solid phase, parameters such as particle size, surface charge and the formation of amorphous complexes needs to be considered. Some metal sulfide precipitates are too small for solid liquid separation by gravity settling. In such processes, fine particles may be generated in the system which makes the settling of metal sulfide particles difficult.

Billon et al. (2001) showed that metals can be adsorbed or co-precipitated with metal oxides/sulfides in the system apart from a direct reaction with sulfide. For instance, Ni can co-precipitate with Mn (Zwolsman et al., 1993). ZnS formation begins with the formation of soluble Zn_3S_3 or $Zn_4S_6^{4-}$, then the crystal formation proceeds further with other anions and ends with insoluble $ZnS_{(s)}$ (Lewis, 2010). Cu is the easiest metal to react with sulfide and at the same time the most difficult to study due to its high compatibility with sulfidic reduction (Lewis, 2010; Luther et al., 1996). NiS may prevail as polymorphs (Ni_xS_y) in the system (Bijmans et al., 2009). Bijmans et al. (2009) reported that during selective Ni precipitation in a gas-lift SRB bioreactor, NiS can be present as insoluble compound such as NiS (millerite), Ni_4S_3, Ni_3S_4 (polydymite) and also as a soluble complex.

Excess sulfide in the system (more than 1:1 ratio of Ni:S^{2-}) can affect the solubility of NiS. Lewis and van Hille (2006) observed that in a fluidized bed reactor with Ni and Co, the metal:sulfide ratio of 1:1 can stabilize metal sulfide precipitates, while a ratio more than 1:1.2 can redissolve both metals and form polysulfide complexes, especially for Ni. The remaining sulfide concentration in the IFBs, which is lower than the MINTEQ predicted values, may thus be due to the generation of polysulfides (Bekmezci et al., 2011; Cao et al., 2009).

5.4.2.2 The presence of organic compounds

The residual COD in the IFB can also affect the HM removal efficiencies. Organic compounds have functional groups which can bind to metal ions. In many physico-chemical treatment processes such as adsorption, the carboxylic group (-COOH) has shown to be a very potential functional group that tends to attract metal ions (Ajmal et al., 2000; Guo et al., 2008; Schiewer and Iqbal, 2010). In this study, the remaining organic compounds are volatile fatty acids (VFA) like acetate, which is a degradation product of the e-donor ethanol (Liamleam and Annachhatre, 2007) present as remaining COD in the effluent. Organic compounds such as acetate or EDTA have the capability to bind the metal ions, especially Zn, which competes with sulfide precipitation (Esposito et al., 2006). There are many reports regarding this phenomenon which occur in nature as well. Yu et al. (2001) and Peng et al. (2009) have reported the interaction between HMs and organics in river sediments leading to the formation of complexes which reduces their mobility in the aqueous phase.

Exopolymeric substances (EPS) may also play an important role by interacting with HMs due to their negative charge. EPS, composed of polysaccharides, phospholipids and proteins, are known for their capability to enhance the cell adhesion during biofilm formation (Stadler et al., 2010). However, the exact properties and composition of the EPS are quite difficult to predict since they can vary based on factors such as hydrodynamics, nutrient availability and microbial community distribution (Ruiz et al., 2008). Ledin (2000) and Comte et al. (2006) described that the mechanisms of EPS-metal binding involves physical adsorption, complexation and ion exchange, including precipitation. Barber and Stuckey (2000) reported that apart from sulfide precipitation, complexation with microbial soluble products could be another possible reason for reducing bioavailability of trace metals in bioreactors. Yin et al. (2013) reported a similar observation in their study using EPS extracts from *Aspergillus fumigatus* for Cd, Cu and Pb removal. EPS from the SRB usually exhibit high affinity for metal ion binding, which is related to metal corrosion (Beech et al., 1999). Wang et al. (2014) also reported that the EPS can adsorb metal ions from aqueous solution, thereby improving the removal efficiency. On the contrary, Villa-Gomez et al. (2011) report that EPS has no effect on the metal removal in an IFB reactor.

Concerning the effect of pH on HM adsorption to EPS, Comte et al. (2008) studied the adsorption of Cd, Cu and Pb onto EPS, and reported that low pH (pH 4.0) hinders the adsorption capacity of EPS, mainly caused by an increase in the number of inactive binding sites at the EPS surface and metal solubility. Salehizadeh and Shojaosadati (2003) also reported that the pH is one of the parameters that affects Cu, Pb and Zn removal using polysaccharides produced from *Bacillus firmus*. Therefore, in SRB reactors where physico-chemical and biochemical reactions govern the efficiency of the process, the pH still plays a major role in the metal removal efficiency. Further studies are required to understand the influence of other electron donors such as lactate, pyruvate or acetate in order to elucidate the SR activities and the kinetics of the bioprocess. By characterizing the functional groups present in EPS, it would be possible to elucidate the mechanisms of biologically induced precipitation in an IFB.

5.5 Conclusions

- The Cu and Zn removal efficiency in an IFB bioreactor were high (> 90% for both metals at both pH 7.0). However, the Ni removal efficiency was affected by the presence of other metal ions (Cu and Zn) in the influent, residual sulfide and EPS produced by the biomass.
- The initial metal concentration and pH played a major role in affecting the reactor performance in terms of COD, sulfate and HM removal efficiency. The presence of Ni (25 mg/L) drastically reduced the SR performance of the IFB, presumably due to toxic effects of Ni to the biofilm.

References

Ajmal, M., Rao, R.A.K., Ahmad, R. and Ahmad, J., 2000. Adsorption studies on *Citrus reticulata* (fruit peel of orange): removal and recovery of Ni (II) from electroplating wastewater. J. Hazard. Mater., 79(1-2): 117-131.

APHA, 1999. *Standard Methods for the Examination of Water and Wastewater. -- 20th Edition/ Prepared and Published Jointly by American Public Health Association, American Water Works Association, Water Environment Federation.* AWWA.

Beech, I.B., Zinkevich, V., Tapper, R., Gubner, R. and Avci, R., 1999. Study of the interaction of sulphate-reducing bacteria exopolymers with iron using X-ray photoelectron spectroscopy and time-of-flight secondary ionisation mass spectrometry. J. Microbiol. Methods, 36(1-2): 3-10.

Bekmezci, O.K., Ucar, D., Kaksonen, A.H. and Sahinkaya, E., 2011. Sulfidogenic biotreatment of synthetic acid mine drainage and sulfide oxidation in anaerobic baffled reactor. J. Hazard. Mater., 189(3): 670-676.

Billon, G., Ouddane, B., Laureyns, J. and Boughriet, A., 2001. Chemistry of metal sulfides in anoxic sediments. PCCP, 3(17): 3586-3592.

Cabrera, G., Pérez, R., Gomez, J., Abalos, A. and Cantero, D., 2006. Toxic effects of dissolved heavy metals on *Desulfovibrio vulgaris* and *Desulfovibrio* sp. strains. J. Hazard. Mater., 135(1): 40-46.

Cao, J., Zhang, G., Mao, Z., Fang, Z. and Yang, C., 2009. Precipitation of valuable metals from bioleaching solution by biogenic sulfides. Miner. Eng., 22(3): 289-295.

Comte, S., Guibaud, G. and Baudu, M., 2006. Biosorption properties of extracellular polymeric substances (EPS) resulting from activated sludge according to their type: Soluble or bound. Process Biochem., 41(4): 815-823.

Comte, S., Guibaud, G. and Baudu, M., 2008. Biosorption properties of extracellular polymeric substances (EPS) towards Cd, Cu and Pb for different pH values. J. Hazard. Mater., 151(1): 185-193.

Cord-Ruwisch, R., 1985. A quick method for the determination of dissolved and precipitated sulfides in cultures of sulfate-reducing bacteria. J. Microbiol. Methods, 4(1): 33-36.

Du Laing, G., Rinklebe, J., Vandecasteele, B., Meers, E. and Tack, F.M.G., 2009. Trace metal behaviour in estuarine and riverine floodplain soils and sediments: a review. Sci. Total Environ., 407(13): 3972-3985.

Eaton, A.D., Franson, M.A.H., Association, A.P.H., Association, A.W.W. and Federation, W.E., 2005. *Standard Methods for the Examination of Water & Wastewater*. American Public Health Association, Washington D.C., the United States.

Esposito, G., Veeken, A., Weijma, J. and Lens, P.N.L., 2006. Use of biogenic sulfide for ZnS precipitation. Sep. Purif. Technol., 51(1): 31-39.

Fu, F. and Wang, Q., 2011. Removal of heavy metal ions from wastewaters: A review. J. Environ. Manage., 92(3): 407-418.

Gallegos-Garcia, M., Celis, L.B., Rangel-Méndez, R. and Razo-Flores, E., 2009. Precipitation and recovery of metal sulfides from metal containing acidic wastewater in a sulfidogenic down-flow fluidized bed reactor. Biotechnol. Bioeng., 102(1): 91-99.

Guo, X., Zhang, S. and Shan, X.-q., 2008. Adsorption of metal ions on lignin. J. Hazard. Mater., 151(1): 134-142.

Jong, T. and Parry, D.L., 2003. Removal of sulfate and heavy metals by sulfate reducing bacteria in short-term bench scale upflow anaerobic packed bed reactor runs. Water Res., 37(14): 3379-3389.

Jong, T. and Parry, D.L., 2006. Microbial sulfate reduction under sequentially acidic conditions in an upflow anaerobic packed bed bioreactor. Water Res., 40(13): 2561-2571.

Kiran, B. and Thanasekaran, K., 2011. Copper biosorption on Lyngbya putealis: Application of response surface methodology (RSM). Int. Biodeterior. Biodegradation, 65(6): 840-845.

Ledin, M., 2000. Accumulation of metals by microorganisms - processes and importance for soil systems. Earth-Sci. Rev., 51(1-4): 1-31.

Lewis, A.E., 2010. Review of metal sulphide precipitation. Hydrometallurgy, 104(2): 222-234.

Liamleam, W. and Annachhatre, A.P., 2007. Electron donors for biological sulfate reduction. Biotechnol. Adv., 25(5): 452-463.

Luther, G.W., Rickard, D.T., Theberge, S. and Olroyd, A., 1996. Determination of metal (bi) sulfide stability constants of Mn^{2+}, Fe^{2+}, Co^{2+}, Ni^{2+}, Cu^{2+}, and Zn^{2+} by voltammetric methods. Environ. Sci. Technol., 30(2): 671-679.

Neculita, C.-M., Zagury, G.J. and Bussière, B., 2007. Passive treatment of acid mine drainage in bioreactors using sulfate-reducing bacteria. J. Environ. Qual., 36(1): 1-16.

Peng, J.-F., Song, Y.-H., Yuan, P., Cui, X.-Y. and Qiu, G.-L., 2009. The remediation of heavy metals contaminated sediment. J. Hazard. Mater., 161(2-3): 633-640.

Peters, R.W., YOUNG, K. and Chang, T.-K., 1984. Heavy metal crystallization kinetics in an MSMPR crystallizer employing sulfide precipitation, AIChE Symp. Ser. American Institute of Chemical Engineers, pp. 55-75.

Ruiz, L.M., Valenzuela, S., Castro, M., Gonzalez, A., Frezza, M., Soulère, L., Rohwerder, T., Queneau, Y., Doutheau, A., Sand, W., Jerez, C.A. and Guiliani, N., 2008. AHL communication is a widespread phenomenon in biomining bacteria and seems to be involved in mineral-adhesion efficiency. Hydrometallurgy, 94(1-4): 133-137.

Salehizadeh, H. and Shojaosadati, S.A., 2003. Removal of metal ions from aqueous solution by polysaccharide produced from *Bacillus firmus*. Water Res., 37(17): 4231-4235.

Sawyer, C., McCarty, P. and Parkin, G., 2003. *Chemistry for Environmental Engineering and Science*. McGraw-Hill Education, the United States.

Schiewer, S. and Iqbal, M., 2010. The role of pectin in Cd binding by orange peel biosorbents: A comparison of peels, depectinated peels and pectic acid. J. Hazard. Mater., 177(1-3): 899-907.

Stadler, R., Wei, L., Fürbeth, W., Grooters, M. and Kuklinski, A., 2010. Influence of bacterial exopolymers on cell adhesion of *Desulfovibrio vulgaris* on high alloyed steel: Corrosion inhibition by extracellular polymeric substances (EPS). Mater. Corros., 61(12): 1008-1016.

Villa-Gomez, D., Ababneh, H., Papirio, S., Rousseau, D.P.L. and Lens, P.N.L., 2011. Effect of sulfide concentration on the location of the metal precipitates in inversed fluidized bed reactors. J. Hazard. Mater., 192(1): 200-207.

Villa-Gomez, D.K., Cassidy, J., Keesman, K.J., Sampaio, R. and Lens, P.N.L., 2014. Sulfide response analysis for sulfide control using a pS electrode in sulfate reducing bioreactors. Water Res., 50(0): 48-58.

Wan Ngah, W.S. and Hanafiah, M.A.K.M., 2008. Removal of heavy metal ions from wastewater by chemically modified plant wastes as adsorbents: a review. Bioresour. Technol., 99(10): 3935-3948.

Wang, J., Li, Q., Li, M.-M., Chen, T.-H., Zhou, Y.-F. and Yue, Z.-B., 2014. Competitive adsorption of heavy metal by extracellular polymeric substances (EPS) extracted from sulfate reducing bacteria. Bioresour. Technol., 163: 374-376.

Wang, S. and Peng, Y., 2010. Natural zeolites as effective adsorbents in water and wastewater treatment. Chem. Eng. J., 156(1): 11-24.

Yin, Y., Hu, Y. and Xiong, F., 2013. Biosorption properties of Cd (II), Pb (II), and Cu (II) of extracellular polymeric substances (EPS) extracted from *Aspergillus fumigatus* and determined by polarographic method. Environ. Monit. Assess., 185(8): 6713-6718.

Yu, K.-C., Tsai, L.-J., Chen, S.-H. and Ho, S.-T., 2001. Chemical binding of heavy metals in anoxic river sediments. Water Res., 35(17): 4086-4094.

Zwolsman, J.J.G., Berger, G.W. and Van Eck, G.T.M., 1993. Sediment accumulation rates, historical input, postdepositional mobility and retention of major elements and trace metals in salt marsh sediments of the Scheldt estuary, SW Netherlands. Mar. Chem., 44(1): 73-94.

Chapter 6 Effects of Process Parameter on Metal Yield in E-Waste Leachate

The modified version of this chapter was submitted for publication as:
Janyasuthiwong S., Ugas R., Rene E.R., Alessandra C., Esposito G. and Lens P.N.L. (2015). Effect of operational parameters on the leaching efficiency and recovery of heavy metals from computer printed circuit boards. *J. Chem. Technol. Biotechnol, Accepted*
DOI: 10.1002/jctb.4798

Summary

Electronic waste (e-waste) is a recent environmental problem worldwide, because it is bulky in size causing solid waste disposal problems, especially due to the release of toxic heavy metals into the environment. On the other hand, metal scarcity is another global issue with e-waste, making it a prominent candidate for metal recovery. In this study, desktop computer printed circuit boards (PCBs) were evaluated as a potential source of Cu, Ni and Zn recovery using a hydrometallurgical method. The process parameters for metal leaching i.e. different particle sizes (< 0.5, 0.5-1.0, 1.0-5.0 and 5.0-9.5 mm), leaching agents (HNO_3, H_2SO_4 and C_2H_5ONa), contact time (18 h), liquid to solid (L/S) ratio (10, 20 and 40), and agitation speed (0, 100 and 200) were evaluated in batch systems. At the following optimal conditions of 1M HNO_3, L/S ratio of 20 with 200 rpm agitation speed using 0.5-1.0 mm particle size, the concentrations of Cu, Ni and Zn leached out from the PCBs were 450, 8 and 4 mg/g PCBs, respectively. Continuous leaching in an up-flow leaching column using the optimized parameters, coupled to sulfide precipitation yielded > 90% Cu recovery (0.48 g/g PCBs).

6.1 Introduction

The improvement in telecommunication technologies and the availability of sophisticated electrical appliances has not only increased the volume of electronics produced, but has also decreased the life time of these equipments. This results in large amounts of these devices being disposed off as electronic waste (e-waste), especially at landfill sites. On a global scale, e-waste generation is growing by about 40 million tons a year. According to the UNEP and UNU report, China generates 64% of the world's e-waste, followed by India (13%) and Brazil (11%), and this is expected to increase by at least 2 to 8-fold by the year 2020 (Schluep, 2009). E-waste is heterogeneous by nature because it contains different elements, product sizes and weights, varying from < 100 g up to > 80 kg (Huisman et al., 2008). Concerning its material composition, modern electronics can contain up to ~60 different elements (Schluep, 2009). Materials generally include all types of metals, glass and plastics which usually have a positive value in the market (Kang and Schoenung, 2005; Realff et al., 2004).

Several reports have found that the pollution level in the surrounding areas of landfills are heavily contaminated with heavy metals, dioxins and polycyclic aromatic hydrocarbons (PAHs) (Ha et al., 2009; Shen et al., 2009; Wang et al., 2005). This pollution is not limited to the environment, but includes also living organisms. Guo et al. (2010) reported that the placenta of human living nearby e-waste recycling sites has high concentrations of Pb, Cd, Cr and Ni. In addition, apart from improper management of the waste disposal sites, improper recycling practices such as aqua regia leaching and open-air burning of e-waste also accelerate their environmental and health impacts (Robinson, 2009; Sepúlveda et al., 2010).

The increasing scarcity of natural resources and the rising cost of mining new (raw) materials have recently created a paradigm shift of turning e-waste into a secondary metal source (Ba et al., 2009; Widmer et al., 2005). Although the burden on the environment and health is still high, the incentive of metal recovery can still yield a good profit since resource scarcity has become one of the recent global issues next to global warming and energy deficiency. Particularly in developing countries, e-waste has a significant role to play in the recycling sector even though it is not necessarily generated in those countries. Many researchers have shifted their focus towards resource recovery from wastes, in which e-waste ranks as a top candidate for metal recovery (Cui and Zhang, 2008).

Hydrometallurgical methods leach the metals out in the aqueous phase. These methods are advantageous over pyrometallurgical methods since they are done at milder conditions (lower temperature) and generate less air emissions. Several technologies have been recommended to recover the metals from the aqueous phase (leachate) such as adsorption, electroplating and precipitation. Sulfide precipitation is one of the methods with a high efficiency and the generated product, *i.e.* metal sulfides, is easy to handle (Lewis, 2010). Another advantage is that the use of biogenic sulfide reduces costs of chemicals in the recovery phase (Jong and Parry, 2003).

In order to optimize the metal recovery from e-waste, there are several process parameters that need to be optimized. In the present study, printed circuit boards (PCBs) of personal computers were used as model e-waste. The effect of particle size, agitation speed, liquid to solid (L/S) ratio, leaching agent and temperature on the metal leaching efficiency of PCBs was investigated in batch systems. Moreover, a continuously operated up-flow leaching column was also tested to ascertain the leaching efficiency of metals from PCBs under the optimized conditions obtained from the batch experiments.

6.2 Material and methods

6.2.1 Experimental material and chemical

6.2.1.1 E-waste material

The desktop computers were collected from the IT department of UNESCO-IHE (Delft, the Netherlands). They were disassembled and only the printed circuit boards (PCBs) were retained for further investigation. The PCBs were crushed using a crusher at TU Berlin (Germany) and sieved into 4 particle size ranges ($d < 0.5$, 0.5-1.0, 1.0-5.0 and 5.0-9.5 mm) as shown in Figure 6.1 with a sieving machine (Fritsch analysette 3 PRO).

Figure 6.1 The different particle size classes of PCBs used in this study

6.2.1.2 Chemical and reagents

The chemicals used in the leaching study were sulfuric acid (H_2SO_4), nitric acid (HNO_3), and sodium acetate (CH_3COONa). For sulfide precipitation, $Na_2S \cdot 9H_2O$ was used as the source of sulfide (Cord-Ruwisch, 1985). The synthetic medium used for biogenic production of sulfide consisted of NH_4Cl 300 mg/L, $MgCl_2 \cdot 6\ H_2O$ 120 mg/L, $KHPO_4$ 200 mg/L, KCl 250 mg/L, $CaCl_2 \cdot 2\ H_2O$ 15 mg/L and yeast extract 20 mg/L, respectively. All chemicals used in this study were of analytical grade.

6.2.2 PCBs characterization

The sieved materials characterized for both physical (magnetic separation and burnable content) and chemical parameters (total metal concentration). For magnetic separation, the crushed PCBs of a size of < 0.5 and 0.5-1.0 mm were used. The PCBs were placed randomly on an analytical balance (Sartorius RC 210D) and weighed initially before they were contacted with a commercial rectangular magnet (~100 gauss). Then, the remaining part on the container was weighed for the non-ferrous part (n=30). The burnable fraction was estimated by placing the material into a muffle furnace at different temperatures (500, 700 and 900 °C) for several treatment durations (15, 30 and 60 min), in a similar way as described by Havlik et al. (2011; 2010). Total metal analysis was done using aqua regia (1 HNO_3: 3 HCl *v/v*) as a leaching agent. The materials were soaked in the leaching agent for 24 h with a L/S ratio of 20.

6.2.3 Metal leaching experiments

6.2.3.1 Batch experiments

Batch experiments were conducted to ascertain the optimal value of each process parameter investigated, *i.e.*, particle size, agitation speed, liquid to solid (L/S) ratio, leaching agent and temperature, while the contact time was set at 18 h according to the toxicity characteristic leaching procedure (TCLP) (USEPA, 2000). The ranges of these process parameters are shown in Table 6.1. Experiments were done in triplicate in 115 mL serum bottles fitted with a septum, the bottles were placed on a platform shaker (Innova 2100).

Table 6.1 Process parameters used in the PCBs batch leaching experiments

Parameters	Conditions
Working vessel volume (mL)	115
Working volume (mL)	20
Temperature (± 3 °C)	5.0, 20.0, 30.0, 60.0
Shaking speed (rpm)	0, 100, 200
Time (h)	18
Particles size (mm)	< 0.5, 0.5-1.0, 1.0-5.0, 5.0-9.5
Leaching agents	1.0 M acetate buffer, 1.0 M nitric acid, 1.0 M sulfuric acid

6.2.3.2 Leaching column

Polyvinylchloride (PVC) columns were set up for the continuous experiments. The column was 60.0 mm in diameter (thickness of 1.5 mm), 400 mm in height and has a total operating volume of ~0.85 L (Figure 6.2). The column was operated in up-flow mode and had an overflow system for effluent collection. The leaching agent used in the leaching column was 1.0 and 0.5 M HNO_3, and was fed by a peristaltic pump (Masterflex model 7528-10, pump head model 77201-60). The operating conditions were a L/S ratio of 20, hydraulic retention time (HRT) of 6.0 h (equal to 1 cycle) and an operating temperature of 25 ± 3°C. The particle size used in the study was 0.5-1.0 mm, which was found to give the best leaching efficiency in the batch experiments, and a mixture of all particle sizes with a 1:1:1:1 ratio. The materials were placed into a cup-type vessel made of polystyrene, and the vessel was kept at a distance of ~5.0 cm from the column inlet.

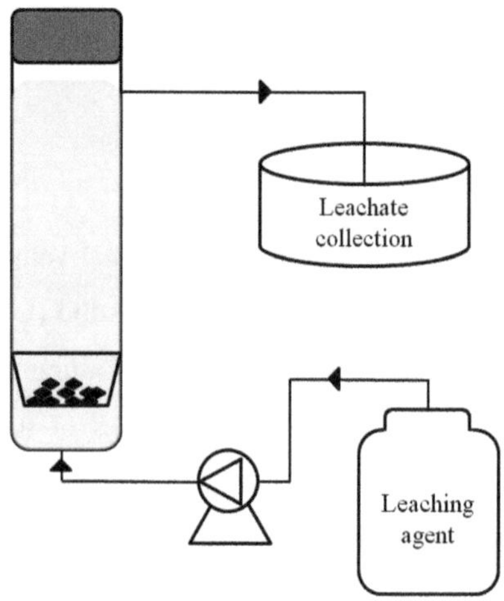

Figure 6.2 Schematic of the continuously operated leaching column for heavy metal dissolution from PCBs

6.2.4 Sulfide precipitation experiments

The leachate used in this experiment was collected from the column experiments with 0.5-1.0 mm and mixed particle size using 1.0 M HNO_3. The sulfide used in this study was produced both chemically, using $Na_2S \cdot 9H_2O$, and biologically from a sulfate reducing batch bioreactor. The inoculum containing sulfate reducing bacteria (SRB) was originally obtained from Biothane Systems International (Delft, the Netherlands) with a VSS of 35.4 g/L. Experiments were performed with Cu:sulfide stoichiometric ratio's of 1:1 and 1:2, respectively.

6.2.5 Analytical methods

The pH was measured with a pH meter (691 Metrohm pH meter) and a SenTix 21 WTW pH electrode. Sulfide was measured by the colorimetric method as described in Cord-Ruwisch

(1985) using a PerkinElmer Lambda 20 Spectrophotometer. Metal concentrations in the leachate were analyzed by flame atomic absorption spectrophotometer (AAS) on a AAnalyst 200 Perkin Elmer (Eaton *et al.*, 2005). For quantifying the total metal concentration in the PCBs, the samples were diluted, acidified with HNO_3 and analyzed by an ICP-MS model XSERIES 2.

6.3 Results

6.3.1 Characterization of PCBs

The ferrous content of the PCBs used in this study was 4.5 and 11.0% for the samples with a particle size of <0.5 and 0.5-1.0 mm, respectively. The range of burnable content of PCBs from all 4 particle size ranges was higher than 25%, but below 35% (Figure 6.3). A temperature of 300 °C was sufficient to get rid of the burnable content of samples having a particle size class <0.5 mm (Figure 6.3a), while the larger particle size required higher temperatures (> 700 °C) due to the non-uniform nature and composition of these particles (Figure 3d).

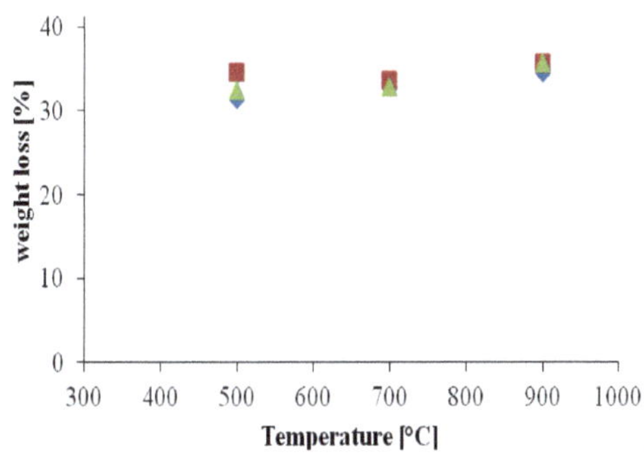

(a)

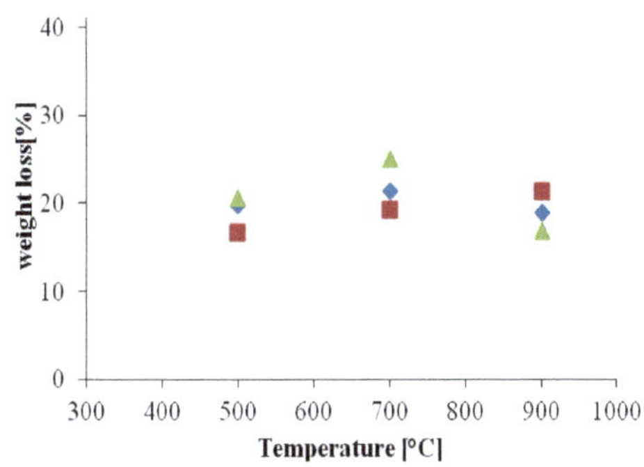

(b)

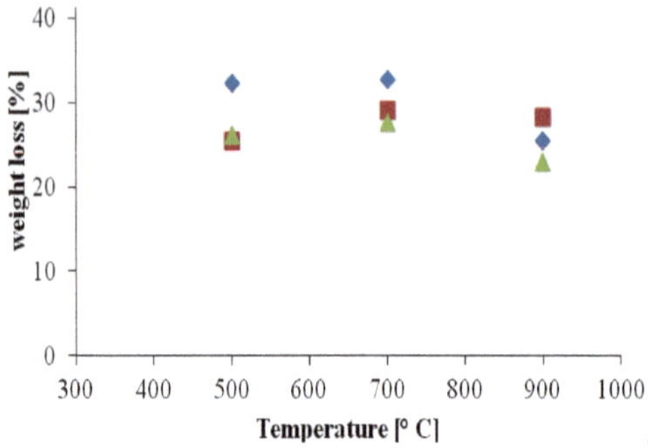

(c)

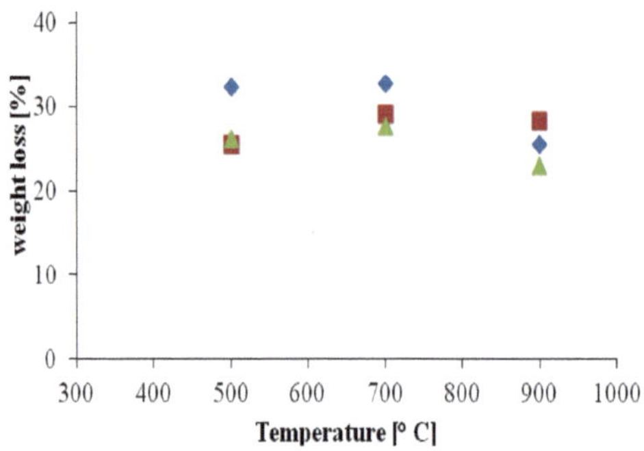

(d)

Figure 6.3 Percentage of burnable content at various temperatures and durations. a) d< 0.5, b) 0.5-1.0, c) 1.0-5.0 and d) 5.0-9.5 mm. (♦) 15 min (■) 30 min and (▲) 40 min

The total metal concentration of the PCBs using aqua regia leaching is shown in Table 6.2. The most abundant metal was Cu with a concentration of 173.1, 352.9 and 237.1 mg/g PCBs for particle size classes < 0.5, 0.5-1.0 and 1.0-5.0 mm, respectively. Pb was found to be the second most abundant metal with 12.54, 22.32 and 28.05 mg/g PCBs for the particle sizes < 0.5, 0.5-1.0 and 1.0-5.0 mm, respectively.

6.3.2 Batch experiments

6.3.2.1 Effect of leaching agents

The optimization of process parameters was done to elucidate the best leaching conditions for maximizing the metal leaching efficiency. Among the different leaching agents screened (1.0 M HNO_3, 1.0 M H_2SO_4 , 1.0 M CH_3COONa) at ambient temperature (~20°C), 1.0 M HNO_3 released most of the heavy metals, while the less effective agent was 1.0 M CH_3COONa. Therefore, CH_3COONa was omitted from further experiments (Figure 6.4). The remaining process parameters were optimized by using 1.0 M HNO_3 and 1.0 M H_2SO_4 as the leaching agents. The metal yield obtained when using 1.0 M HNO_3 was 450, 7 and 9 mg/g PCBs for Cu, Ni and Zn, respectively.

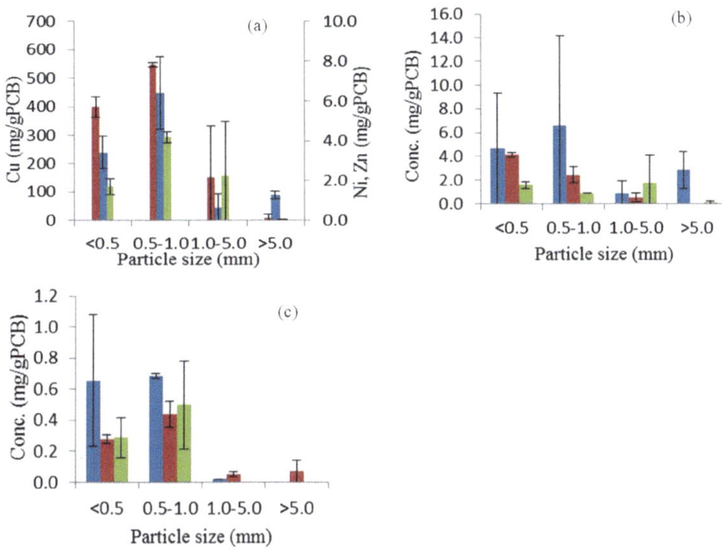

Figure 6.4 Screening of leaching agents using different particle size classes of PCBs at 200 rpm, L/S ratio of 20 and contact time of 18 h. (a) HNO_3, (b) H_2SO_4 and (c) CH_3COONa. (■) Cu, (■) Ni and (■) Zn

6.3.2.2 Effect of agitation speed

To evaluate the effect of mixing on the leaching potential of heavy metals from PCBs, experiments were carried out at ambient temperature using all particle size fractions, a L/S ratio of 20, and at an agitation speed of 100 and 200 rpm. 1.0 M HNO_3 and 1.0 M H_2SO_4 were used as the leaching agents. Among all the tested size fractions, PCBs within the range of 0.5-1.0 mm showed the best results irrespective of the agitation speed (Figure 6.5). The results of leaching with HNO_3, after 18 h, and at a mixing speed of 100 rpm showed that the concentrations of Cu, Ni and Zn in the leachate reached 275.0, 13.0 and 2.5 mg/g PCBs, respectively (Figure 6.5a). The results obtained at an agitation speed of 200 rpm, for 18 h, also showed the highest concentrations for Cu (450 mg/g PCBs), followed by Ni (8 mg/g PCBs) and Zn (4 mg/g PCBs) (Figure 6.5b). Experiments conducted with 1.0 M H_2SO_4, under the same conditions, showed less values for Cu, Ni and Zn dissolution (< 12 mg/g PCBs).

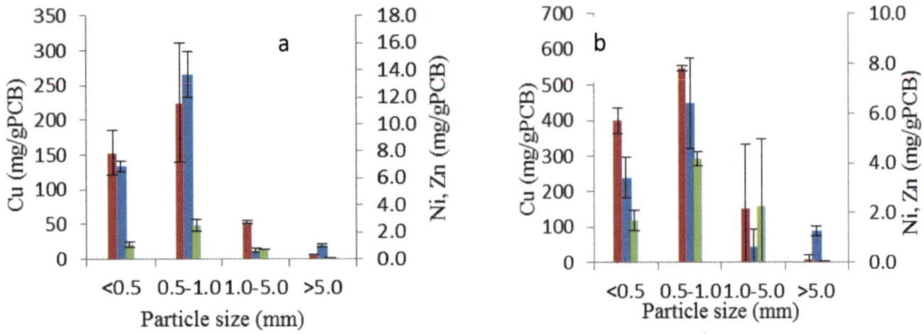

Figure 6.5 Effect of agitation speed on the metal leachability using 1.0 M HNO_3 as the leaching agent abd a L/S ratio of 20. (a) 100 rpm, and (b) 200 rpm. (■) Cu, (■) Ni and (■) Zn

Table 6.2 Total metal concentration in the PCBs at different size classes

	Concentration (mg/g PCBs)											
	Al	Ba	Cu	Cd	Cr	Fe	Ga	Mo	Ni	Pb	Ti	Zn
d<0.5 mm	2.05	3.25	173.1	N.D.	0.13	2.14	N.D.	N.D.	7.19	12.54	N.D.	1.68
0.5-1.0 mm	1.50	2.62	352.9	N.D.	0.16	3.02	N.D.	N.D.	8.15	22.32	N.D.	10.916
1.0-5.0 mm	0.57	0.68	237.1	N.D.	N.D.	0.79	N.D.	N.D.	3.89	28.05	N.D.	13.27

Note: N.D. = lower than detection limit

6.3.2.3 Effect of L/S ratio

Batch experiments were conducted with 1.0 M HNO_3 and 1.0 M H_2SO_4, at an agitation speed of 200 rpm, for 18 h, at different particle size fractions and at different L/S ratio's of 10 and 40 (Figure 6.6). These results were compared with the previously obtained concentrations of Cu, Ni and Zn when experiments were performed at a L/S ratio of 20 (Figure 6.5b). It is quite evident from these plots that there was not a significant difference in the leaching characteristics at a L/S ratio of 10 and 20, wherein the concentration of Cu, Ni and Zn reached nearly similar values: ~450.0, 9.0 and 3.0 mg/g PCBs. On the other hand, the metal concentration at a L/S ratio of 40 reached values of 175.0, 9.0 and 5.0 mg/g PCBs for Cu, Ni and Zn, respectively. It was also evident that HNO_3 is more effective than H_2SO_4, irrespective of the L/S ratio's tested (data not shown).

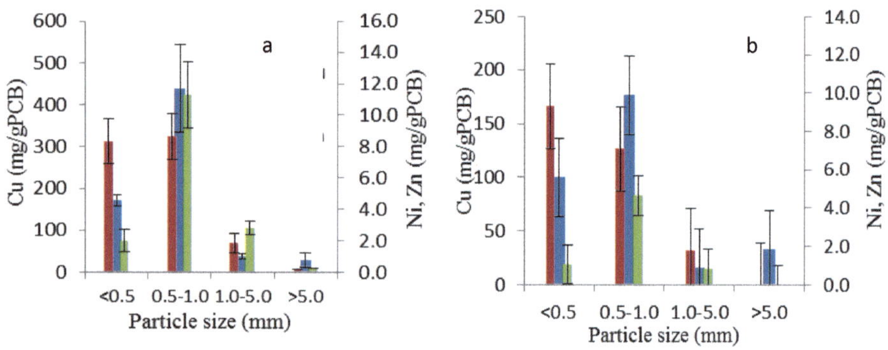

Figure 6.6 Effect of L/S ratio on the metal leachability using 1.0 M HNO_3 as the leaching agent and at an agitation speed of 200 rpm. (a) L/S ratio: 10 and (b) L/S ratio: 40. (■) Cu, (■) Ni and (■) Zn

6.3.2.4 Effect of temperature

Batch experiments were performed at three different temperatures (5, 30 and 60°C) at a L/S ratio of 20, shaking speed of 200 rpm for 18 h using 1.0 M HNO_3 and 1.0 M H_2SO_4 as the leaching agents (Figure 6.7). As previously done, the results were compared with the experimental results obtained at 20°C (Figure 6.5b). At 5°C, the metal dissolution was lower, the Cu concentration reached 175 mg/g PCBs, while the Ni and Zn concentration reached a maximum concentration of 8 and 5 mg/g PCBs, respectively. It can be equated from these figures that temperatures <20°C do not show any significant improvement in the metals leachability. Cu concentrations after leaching with 1.0 M HNO_3 at 30 and 60°C reached maximum values of ~435 and 461 mg/g PCBs, respectively. In the case of Ni and Zn, they reached the highest concentrations of 11.5 and 20.0 mg/g PCBs at 60 and 30°C, respectively. As observed earlier, when leaching experiments were performed with H_2SO_4, the concentrations of individual metals did not exceed 16 mg/g PCBs, irrespective of the operating temperatures (data not shown).

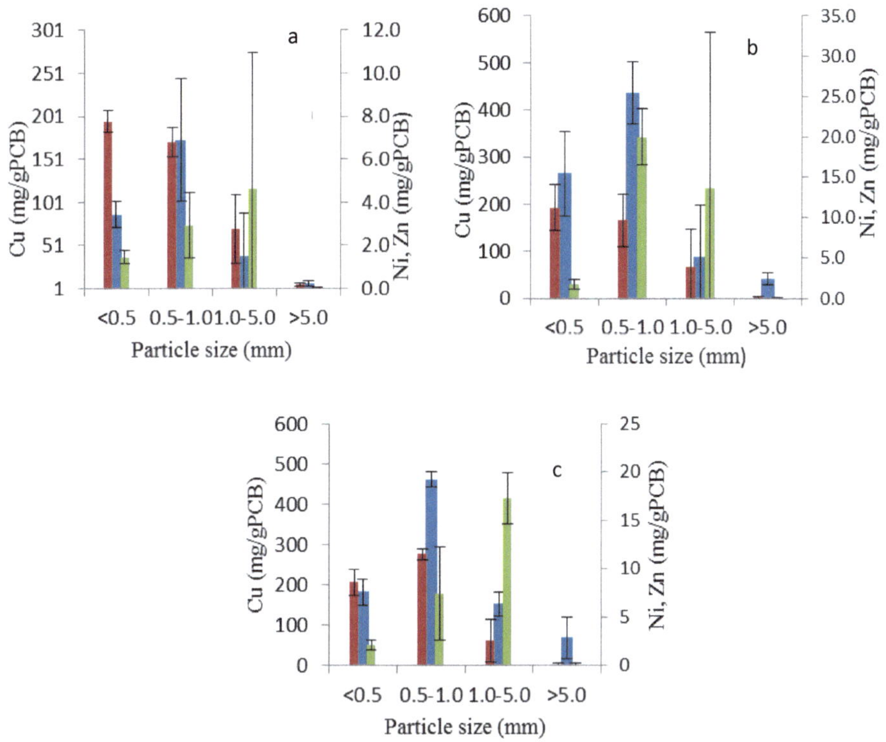

Figure 6.7 Effect of temperature on the metal leachability using 1.0 M HNO_3 as the leaching agent, a L/S ratio of 20 and an agitation speed of 200 rpm. (a) 5 °C, (b) 30 °C and (c) 60 °C. (■) Cu, (■) Ni and (■) Zn

The process parameters that yielded the highest leachability of the metals can be summarized as follows: leaching agent - 1.0 M HNO_3, agitation speed - 200 rpm, L/S ratio - 20, temperatures - 30 and 60°C, particle size - 0.5-1.0 mm and contact time - 18 h. The maximum metal yield in the leachate achieved at these optimal conditions was in the range of 435-461, 9.7-11.5 and 7.4-20 mg/g PCBs for Cu, Ni and Zn, respectively.

6.3.3 Continuous leaching column

Experiments in order to simulate a continuously operated process were also conducted in a leaching column. The maximum Cu concentration was 70.2 and 33.0 mg/g PCBs for the 0.5-1.0 mm and the mixed particle size, respectively (Figure 6.8). The maximum concentration of Ni was reached faster in the second and third cycle, at a concentration of 2.8 and 0.66 mg/g PCBs, respectively. For Zn, the maximum concentration was 0.8 mg/g PCBs in cycle 6 (mixed particle size), while the maximum Zn concentration of 0.5 mg/g PCBs was leached out in cycle 3 of the size class 0.5-1.0 mm, but the concentration of Zn remained stable until 100 h. The concentration of free metal investigated in the leachate of the mixed particle size was quite stable after 78 h, therefore, the experiment was stopped at 82 h (14 cycles).

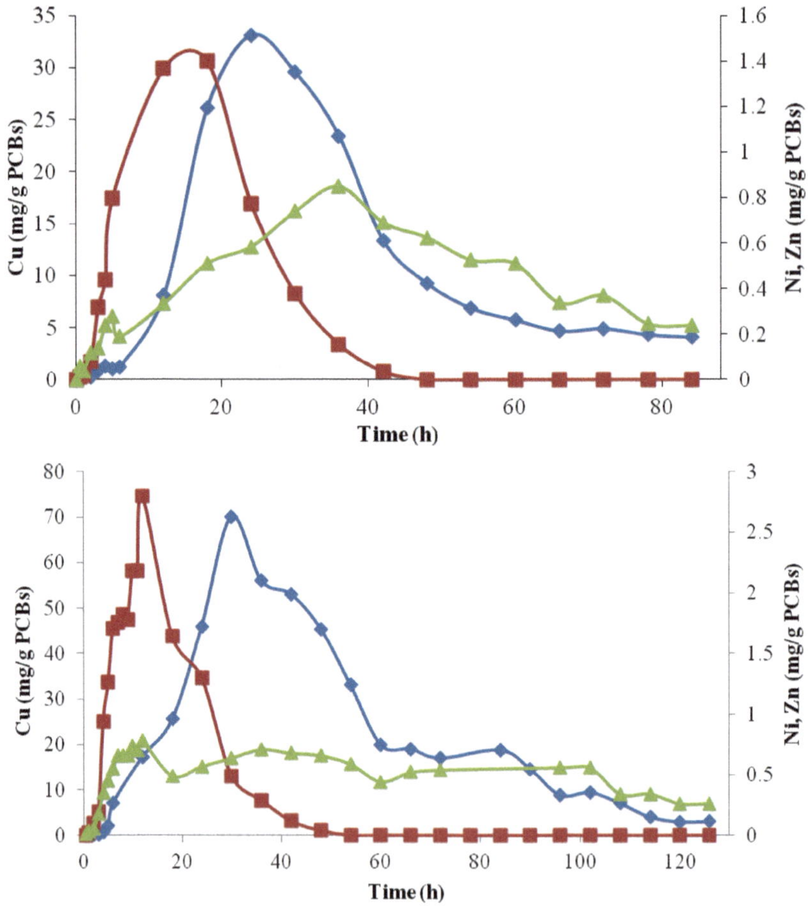

Figure 6.8 Metal concentration in the leachate during continuous experiments using 1.0M HNO_3, (a) 0.5-1.0 mm and (b) mixed particle size. (♦) Cu, (■) Ni and (▲) Zn

6.3.4 Sulfide precipitation

Sulfide precipitation is one of the easiest methods which can achieve fast and cheap metal recovery, especially Cu. The removal efficiencies achieved for Cu, Ni and Zn at a S^{2-}:Cu ratio of 2:1 were 98, 63 and 89% for the particles size class 0.5-1.0 mm, while 98, 49 and 45% was achieved for the mixed particle size sample, respectively. At the S^{2-}:Cu stoichiometric ratio of 1:1, the metal recovery efficiencies were 95, 31 and 44% (0.5-1.0 mm) and 86, 51 and 53% (mixed particle size) for Cu, Ni and Zn, respectively (Figure 6.9).

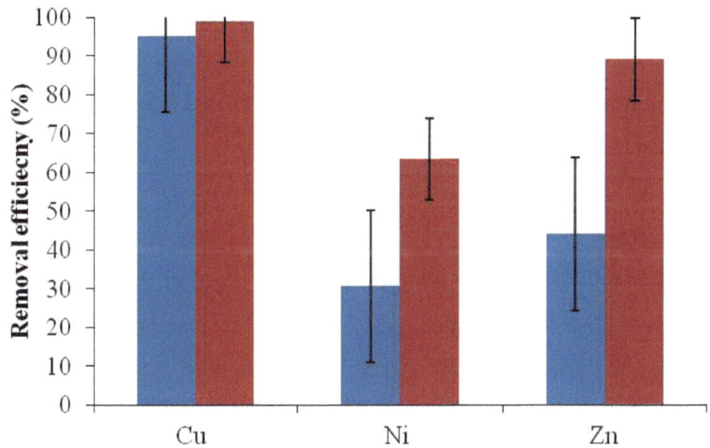

(a)

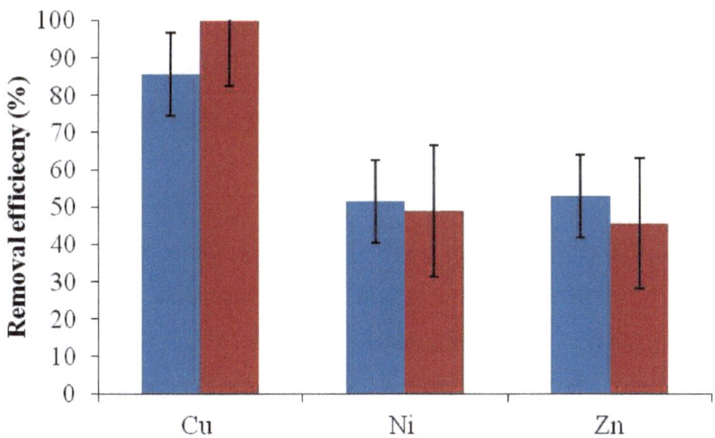

(b)

Figure 6.9 Metal recovery efficiencies from the effluent of the leaching column using, (a) 0.5-1.0 mm size class and (b) mixed particle size class. ( ) S^{2-}:Cu ratio1:1 and ( ) S^{2-}:Cu ratio2:1

6.4. Discussion

6.4.1 PCBs composition

From the characterization of the PCBs materials, the ferrous content is around 4.5 and 11.0%, while the non-ferrous content is 95.5 and 89.0% for the size classes < 0.5 and 0.5-1.0 mm, respectively. The commercial iron magnet used in the study has a magnetic flux density of ~100 gauss which may lack the attracting force for the larger size class PCBs. Oh et al. (2003) reported that the Cu, Ni and Zn contents in the non-magnetic fraction after applying a magnetic flux density of 8000 gauss were 489, 1 and 10 mg/g (48.9, 0.1 and 1%), respectively, among which Cu ranks the first. The Pb concentration is ranked second in that study; however in most cases, the concentration levels of lead in the leachate is mainly related towards environmental and health impact, rather than market value due to its high toxicity (Needleman, 2004).

6.4.2 Metal leachability

Concerning metal composition, Cu is the most abundant metal in the PCBs while there are also other elements in moderate and minor concentrations that could potentially leach out, i.e., Ni, Zn, and Pb (Havlik et al., 2010; Kim et al., 2011; Lee et al., 1997; Mecucci and Scott, 2002; Yang et al., 2011). The best leaching conditions elucidated from this study were 1.0 M HNO_3 as the leaching agent, temperatures of 30 and 60 °C, agitation speed of 200 rpm, L/S ratio of 20 and a particle size class of 0.5-1.0 mm. From Table 2, it is quite apparent that the concentration of copper is the highest, ~352.9 mg/g PCBs, for the particle size class 0.5-1.0 mm (the best condition obtained in the batch experiments), which is a very good indication for Cu recovery.

Mecuci and Scott (2002) reported that HNO_3 is effective in the selective leaching of Cu and Pb from PCBs. Moreover, they also reported that temperature plays a major role in facilitating the migration of the leaching agent into the inner layer of the PCBs, since higher temperatures can cause a considerable drift in the viscosity of the epoxy coating of PCBs. For the particle size, the major reason which made a smaller size class yield better results may come from the increased contact area and better permeability for leaching agents. Koyama et al. (2006) reported that the leaching of Cu can be limited by the conformal coating layers of the PCBs that hinder the diffusion of leaching agents. Yang et al. (2011) also reported that the majority of Cu is present in the middle layer of the PCBs, placed above the laminate. This finding is also supported by many patents on PCBs manufacturing processes, which are always done in multilayer.

Concerning the L/S ratio, this parameter may play a major role in determining the saturation levels of the metal in the leachate. Kinohshita et al. (2003) reported that increasing the L/S ratio resulted in lower Cu and Ni concentrations in the leachate, which is similar to the results obtained in this study (Figure 6). Although the L/S ratio of 10 (Figure 6a) did not show much different metal leaching yields compared to a L/S ratio of 20 (Figure 5b), the L/S ratio of 20 was preferred due to the difficulty in separating the leachate from the remaining solids at a L/S ratio of 10.

6.4.3 Continuous leaching of PCBs

In the continuous leaching column, the first cycle of the column leaching did not yield sufficiently high leaching efficiencies, but Ni recovery may be feasible in the very first cycle. According to Figure 8, the Ni concentration reaches its highest peak in the second and third cycle for the particle size class 0.5-1.0 mm and mixed particle size, respectively. Metal recovery using sulfide precipitation is one of the better choices due to the fast reaction and stable products after the reaction which can tolerate a wider pH range than hydroxide and carbonate precipitation (Lewis, 2010). More than 90% of Cu can be recovered from the leachate (Figure 6.9).

6.5 Conclusion

This study showed that PCBs can be used as a secondary source for base metals, especially Cu. The optimal process conditions for chemical leaching are: leaching agent - 1.0 M HNO_3, agitation speed - 200 rpm, L/S ratio - 20, temperatures - 30 and 60°C, particle size - 0.5-1.0 mm. For a continuous process, Ni recovery should be done during the 2nd and 3rd cycle for the 0.5-1.0 mm and mixed particle size class, respectively. Zn recovery is more difficult since the leaching rates were stable throughout the experimental period. Cu recovery with sulfide precipitation is very effective (> 90%) at a $Cu:S^{2-}$ ratio of 1:2.

References

Ba, T., Zheng, M., Zhang, B., Liu, W., Xiao, K., Zhang, L., 2009. Estimation and characterization of PCDD/Fs and dioxin-like PCBs from secondary copper and aluminum metallurgies in China. Chemosphere 75, 1173-1178.

Cord-Ruwisch, R., 1985. A quick method for the determination of dissolved and precipitated sulfides in cultures of sulfate-reducing bacteria. J. Microbiol. Methods 4, 33-36.

Cui, J., Zhang, L., 2008. Metallurgical recovery of metals from electronic waste: A review. J. Hazard. Mater. 158, 228-256.

Eaton, A.D., Franson, M.A.H., Association, A.P.H., Association, A.W.W., Federation, W.E., 2005. *Standard Methods for the Examination of Water & Wastewater.* American Public Health Association, Washington D.C., the United States.

Guo, Y., Huo, X., Li, Y., Wu, K., Liu, J., Huang, J., Zheng, G., Xiao, Q., Yang, H., Wang, Y., Chen, A., Xu, X., 2010. Monitoring of lead, cadmium, chromium and nickel in placenta from an e-waste recycling town in China. Sci. Total Environ. 408, 3113-3117.

Ha, N.N., Agusa, T., Ramu, K., Tu, N.P.C., Murata, S., Bulbule, K.A., Parthasaraty, P., Takahashi, S., Subramanian, A., Tanabe, S., 2009. Contamination by trace elements at e-waste recycling sites in Bangalore, India. Chemosphere 76, 9-15.

Havlik, T., Orac, D., Petranikova, M., Miskufova, A., 2011. Hydrometallurgical treatment of used printed circuit boards after thermal treatment. Waste Manage. 31, 1542-1546.

Havlik, T., Orac, D., Petranikova, M., Miskufova, A., Kukurugya, F., Takacova, Z., 2010. Leaching of copper and tin from used printed circuit boards after thermal treatment. J. Hazard. Mater. 183, 866-873.

Huisman, J., Magalini, F., Kuehr, R., Maurer, C., Ogilvie, S., Poll, J., Delgado, C., Artim, E., Szlezak, J., Stevels, A., 2008. review of directive 2002/96 on waste electrical and electronic equipment (WEEE), UNU, Bonn, Germany.

Jong, T., Parry, D.L., 2003. Removal of sulfate and heavy metals by sulfate reducing bacteria in short-term bench scale upflow anaerobic packed bed reactor runs. Water Res. 37, 3379-3389.

Kang, H.-Y., Schoenung, J.M., 2005. Electronic waste recycling: A review of U.S. infrastructure and technology options. Resour. Conserv. Recy. 45, 368-400.

Kim, E.-Y., Kim, M.-S., Lee, J.-C., Jeong, J., Pandey, B.D., 2011. Leaching kinetics of copper from waste printed circuit boards by electro-generated chlorine in HCl solution. Hydrometallurgy 107, 124-132.

Kinoshita, T., Akita, S., Kobayashi, N., Nii, S., Kawaizumi, F., Takahashi, K., 2003. Metal recovery from non-mounted printed wiring boards via hydrometallurgical processing. Hydrometallurgy 69, 73-79.

Koyama, K., Tanaka, M., Lee, J.-C., 2006. Copper leaching behavior from waste printed circuit board in ammoniacal alkaline solution. Mater. Trans. 47, 1788.

Lee, J.-C., Koyanaka, S., Lee, M., Ohya, H., Endoh, S., 1997. Recovery of copper, tin and lead from the spent printed circuit boards (PCBs) by the shape separation method. J. Min. Mater. Process. Inst. Jpn. 113, 357-362.

Lewis, A.E., 2010. Review of metal sulphide precipitation. Hydrometallurgy 104, 222-234.

Mecucci, A., Scott, K., 2002. Leaching and electrochemical recovery of copper, lead and tin from scrap printed circuit boards. J. Chem. Technol. Biotechnol. 77, 449-457.

Needleman, H., 2004. Lead poisoning. Annu. Rev. Med. 55, 209-222.

Oh, C.J., Lee, S.O., Yang, H.S., Ha, T.J., Kim, M.J., 2003. Selective leaching of valuable metals from waste printed circuit boards. J. Air Waste Manag. Assoc. 53, 897-902.

Realff, M.J., Raymond, M., Ammons, J.C., 2004. E-waste: an opportunity. Mater. Today 7, 40-45.

Robinson, B.H., 2009. E-waste: An assessment of global production and environmental impacts. Sci. Total Environ. 408, 183-191.

Schluep, M., 2009. *Recycling-from E-waste to Resources: Sustainable Innovation Technology Transfer Industrial Sector Studies*. UNEP, Berlin, Germany.

Sepúlveda, A., Schluep, M., Renaud, F.G., Streicher, M., Kuehr, R., Hagelüken, C., Gerecke, A.C., 2010. A review of the environmental fate and effects of hazardous substances released from electrical and electronic equipments during recycling: examples from China and India. Environ. Impact. Asses. Rev. 30, 28-41.

Shen, C., Chen, Y., Huang, S., Wang, Z., Yu, C., Qiao, M., Xu, Y., Setty, K., Zhang, J., Zhu, Y., Lin, Q., 2009. Dioxin-like compounds in agricultural soils near e-waste recycling sites from Taizhou area, China: Chemical and bioanalytical characterization. Environ. Int. 35, 50-55.

USEPA, 2000. *Test Methods for Evaluating Solid Waste Method 1311, SW-846*. United States Environmental Protection Agency, Washington DC, the United States.

Wang, D., Cai, Z., Jiang, G., Leung, A., Wong, M.H., Wong, W.K., 2005. Determination of polybrominated diphenyl ethers in soil and sediment from an electronic waste recycling facility. Chemosphere 60, 810-816.

Widmer, R., Oswald-Krapf, H., Sinha-Khetriwal, D., Schnellmann, M., Böni, H., 2005. Global perspectives on e-waste. Environ. Impact. Asses. Rev. 25, 436-458.

Yang, H., Liu, J., Yang, J., 2011. Leaching copper from shredded particles of waste printed circuit boards. J. Hazard. Mater. 187, 393-400.

Chapter 7 General Discussion

7.1 Introduction

Science and technology advancements have increased the metal consumption exponentially and accelerated reservoir depletion (Roberts, 1990; Yellishetty et al., 2011). Majorly, the metal is fed to industries as raw material to produce the product in order to convenience daily human life, such as electronic appliances. The global metal production is approximately around 20 to 25 million tones/year in which Europe, the US and Australia are the major producers (Robinson, 2009). The demand for new electronic products with more luxurious design, and effectiveness has made the life span shorter for these appliances. As a consequence, electronics waste (e-waste) has emerged to be a problem in waste management and pollution control with approximately 4% growth per year (Schwarzer et al., 2005).

However, the productions of these electronic equipments require high quantities of metals. The discarded e-waste still contains metals which can be potentially reused or recovered. Many researchers have started to shift their focus towards recovering valuable resources from wastes, among which e-waste ranks as the top candidate for metal recovery (Cui and Zhang, 2008). One of the feasible methods for metal recovery is using hydrometallurgical method, wherein the metals are leached out in the aqueous phase; while, the other technique is pyrometallurgical. The hydrometallurgical method gives more advantage than the pyrometallurgical method in terms of operating conditions (lower temperature), performance, flexibility in operation and low generation of other end-products (especially air emissions) (Andrews et al., 2000).

7.2 Importance of proper metal recovery from e-waste

The rapid population growth still seems to be an unsolved issue and therefore resource consumption and waste generation keeps increasing. Land availability, resource management and environmental quality are also other associated problems (Cropper and Griffiths, 1994; Deng et al., 2008; Ehrlich and Holdren, 1971; Holdren and Ehrlich, 1974). With the limited space on earth surface, the land area has been divided for living, food production and waste storage/treatment. Therefore, there is a high possibility that all of these activities may overlap each other in some areas which will lead into other parallel problems at the later stage.

As the incentive for valuable metals present in e-waste is very high, this business has gained globally recognition (Widmer et al., 2005). India and China are typical examples of those countries which are facing enormous problems associated with the rise in quantity of e-waste handled, both from their domestic disposal and illegal trading from other countries. Wong et al. (2007) investigated the e-waste recycling site in Guiyu, China as the study case and found that the e-waste in the site were from developed countries. The fault of this scenario may also lay in the unrestricted and uncontrolled policies and management of the receiving countries. As a consequence, this site and the neighborhoods are highly contaminated with heavy metal, dioxin, furans, etc. due to primitive handling practices, improper recycling and management.

Several recent reports have highlighted the environmental issues around the vicinity of e-waste recycling sites (Ha et al., 2009; Shen et al., 2009; Wang et al., 2005). Moreover, these recycling sites which use improper metal recovery techniques such as open-air burning or aqua regia leaching without proper environmental and health management have accelerated severe health problems (Robinson, 2009; Sepúlveda et al., 2010). The level of heavy metals in living organisms in these areas such as rice, vegetable, aquatic species including humans are found to be above the threshold limit of metals toxicity (Fu et al., 2008; Guo et al., 2010; Ha et al., 2009; Luo et al., 2011). This scenario is quite drastic, especially the recipients are the children. Zheng et al. (2008) has found that the metal level in the blood of children living nearby recycling site is higher than those live further away.

However, the recycling and recovery of metal from e-waste if processed properly with good management and monitoring system, will yield more benefits not only in terms of resource regeneration but also waste reduction (Babu et al., 2007). The composition of the e-waste is mainly iron and steel which accounts to nearly 50% of its total weight, while, plastic represents the second abundant (21%) and precious metals and non-ferrous metal ranks in third place (13%) (Widmer et al., 2005).

7.3 Metals in e-waste

As mentioned in the previous section, high amount of metals are consumed in the production of electronics. In this section, some typical examples of metals present in e-waste will be demonstrated.

7.3.1 Base metals

The most abundant metal in e-waste, especially in printed circuit boards (PCBs), is Cu, while there are also other metals like Ni, Zn, and Pb (Havlik et al., 2010; Kim et al., 2011; Lee et al., 1997; Mecucci and Scott, 2002; Yang et al., 2011). These metals can be easily found in the stock supply, but some of them have been predicted to be finished from the stock in few decades (Dodson et al., 2012).

7.3.2 Rare metals

Gallium is one of the fundamental elements found in all electronics (Merian et al., 2004) which is extracted and concentrated from bauxite and also as a byproduct of zinc and aluminum refining (Downs, 1993). The utilization of this element has started to attract many interest after the discovery of its semiconductivity property when combined with the Group 15 elements (Greber, 2000). As a consequence of this development, gallium arsenide (GaAs) and gallium nitride (GaN) are the most common compounds used in electronics (Moskalyk, 2003). Another interesting element found in the e-waste is germanium (Ge) which is mainly found in the fiber optic part.

7.3.3 Precious elements

There are others valuable metals that are present within e-waste such as gold (Au), platinum (Pt) and silver (Ag) (Nnorom and Osibanjo, 2008a; Tuncuk et al., 2012). These elements may be found at moderately to low concentration in the electronic components, but it still poses a good indication for recovery.

7.3.4 Rare earth elements

Rare earth elements (REE) are another interesting group that needs to be considered. This element has started to play a major role in the modification of electronics, especially Pb components, since there are restrictions on the use of certain hazardous substances (Wu and Wong, 2007). Moreover, the interest in REE recovery starts from the supply cut off from China (Du and Graedel, 2011; Reck and Graedel, 2012). Some examples of electronics containing REE are TV screen (Eu), hard disk (Dy), and yttrium iron garnet ferrites (Y) (Castor and Hedrick, 2006).

7.4 Metal removal and recovery technology

7.4.1 Combination system

The major characteristics of these waste steam are low pH and are concentrated with various types of metal which is similar to wastewater like acid mine drainage (AMD). For this scenario, metals are already present in aqueous forms which are easier to treat than those that still remain in the solid state. The fundamental of metal contaminated wastewater treatment technology overlaps with other waste treatment/management technologies. This seems to be a good opportunity for the integration of available technologies which will help to solve several environmental issues at the same time. The proposed full process in this chapter includes: sulfate (SO_4^{2-}), sulfide (S^{2-}) precipitation, adsorption and phytoremediation as shown in Table 7.1 which will be further described in the later sections.

Table 7.1 Summary of the proposed processes and their characteristics

Process	Purpose	Advantages	Cost reduction option
Sulfate precipitation	Metal recovery	Recovers metal selectively (Ba and Pb)	Using wastewater containing sulfate
Sulfide precipitation	Metal recovery	Recovery of metals	Using the wastewater effluent containing sulfide
Post treatment			
Adsorption	Post treatment of the effluent	Recovery of S compound and metals	Using agricultural wastes
Phytoremediation	Post treatment of the effluent	Cheap, environmental friendly, nutrient recovery	Use the species which have economical value

7.4.1.1. Sulfate precipitation

One of the well-known methods for SO_4^{2-} treatment is chemical precipitation. In most case, $BaCl_2$ or $Pb(NO_3)_2$ are used in the process which establishes high removal efficiencies, >90% (Benatti et al., 2009; Hlabela et al., 2007; Maree et al., 2004; Špaldon et al., 2013). On the other hand, this method has already proved to be effective for Ba^{2+} and Pb^{2+} removal.

Advantages: Easy and fast process, less maintenance, easy monitoring and less sophisticated system. This process is also applicable for the separation of Ba^{2+} and Pb^{2+} from other metals.

Disadvantages: The major drawbacks of this process are the toxic compounds that are generated as end-products, especially Pb compounds and the separation of precipitates from the aqueous phase. Ca^{2+} may interfere with the process since it can also precipitate with SO_4^{2}, but it has higher solubility than $BaSO_4$ and $PbSO_4$.

7.4.1.2. Sulfide precipitation

This method is applicable for many metals and in a wider range of pH than hydroxide and carbonate (Lewis, 2010). Cu is the most sensitive metal to this method. Choi et al. (2006) reported that the precipitate of Cu is mainly cupric sulfide which can be easily recycled in photochemical and ceramic processes.

Advantages: Easy and fast process. It can tolerate wide range of pH conditions and it is also possible to do selective metal precipitation. High purity of metals can be obtained, especially as CuS and ZnS (Tabak et al., 2003).

Disadvantages: Separation of precipitates from the aqueous phase is one of the major drawbacks of this process (Mersmann, 1999; Peters et al., 1984), while another drawback is the purity of metal precipitate obtained since some metals can co-precipitate. Moreover, sulfide is considered as a toxic compound which may require post-treatment in case they are generated in excess.

7.4.1.3. Post treatment

Although the major concern in the overall process is to recover the metals, the effluent coming from the process is obligate to effluent standards set by various organizations. The following sections provide details on available technologies that have shown high metal removal efficiencies and can be use as post treatment process.

- *Adsorption:*

Adsorption has proven to be an effective physico-chemical technology for removing heavy metals from wastewater, especially for treating wastewaters with a low metal concentration.

This process can be used to remove sulfide and the remaining metals present in water before ultimate discharge.

- *Phytoremediation*

This process is one of the environmental friendly processes using different plant species. This technology is quite well-known for both aesthetic viewpoint and wastewater treatment technology. It can treat various types of waste stream from organics to heavy metals and it can also recover the nutrients.

7.4.2 Process cost optimization

Cost-benefit is one of the important factors involved in deciding on choosing an appropriate technology. These include cost involved in wastewater transportation, treatment and safe environmental disposal (quality). The lower the treatment costs, the more profitable and attractive is wastewater for further use. There are many approaches available in order to reduce the operational cost such as the use of cheap reagent or material, to use the right amount of reagent and reduce waste generation, reuse the effluent of other process(es), or receive benefit from product/byproduct generated from different processes. In this section, several options for cost reduction in each process will be demonstrated. The proposed full process is shown in Figure 7.1.

- Source of SO_4^{2-}:

To minimize the cost of adding SO_4^{2-} chemically, the SO_4^{2-} in the waste streams is another alternate option since it can cause environmental problem such as hydrogen sulfide formation if left untreated (Lens and Kuenen, 2001). Jang and Townsend (2001) reported that the leachate from construction and debris fines (C&D) contained sulfate at very high levels since the main component of C&D is gypsum ($CaSO_4 \cdot 2H_2O$).

- Source of S^{2-}:

The used of sulfate reducing bacteria (SRB) in metal contaminated wastewater treatment is quite well-known for its efficiency and cost-effectiveness (Bijmans et al., 2011; Kaksonen and Puhakka, 2007). There are many wastewater streams that contains high levels of sulfate, such as AMD, pulp and paper wastewater or construction wastes (Lens and Kuenen, 2001; Lens et al., 2000). In addition to the use of sulfate as the electron acceptor for the process, other sulfur compounds such as thiosulfate can also be used (Alazard et al., 2003; Barrett and Clark, 1987; Jørgensen and Bak, 1991; Pikaar et al., 2013).

- Low cost material as an adsorbent:

Generally, agricultural wastes poses less economical value but generate lot of solid waste problems. However, these materials can be used for metal adsorption due to the inheerent characteristics of these materials such as functional group, surface charge and carbon content (Kamsonlian et al., 2011; Orhan and Büyükgüngör, 1993; Sud et al., 2008). Adsorption process can be done at very low cost if the materials are chosen from agricultural waste that less modification in their physico-chemical properties (Dubey and Gopal, 2007; Kadirvelu et al., 2001; Mohan and Singh, 2002). However, based on literature reports, most of the low-cost agricultural waste adsorbents appear to be freely available. However, some of them demand some pretreatment in order to improve their metal adsorption capacities.

- Process control

Process control is another field of study which uses a computer program to control the action of each process to give the desirable outcome. This innovation has been applied in many fields including bioprocesses for wastewater and waste-gas treatment (Dunn et al., 2005). Process control involves data collection, statistics and modeling to tune the PID (proportional-integral-derivative) parameters for each process of interest, for example pH, alkalinity and gas concentration are variables that are to be tuned in an anaerobic process (Pind et al., 2003). By using the predictions, the input and output will be optimized, while the waste generated will be minimized.

For application in metal precipitation, the program can control the feeding of influent to have proper ratio of e-waste leachate (containing metals) and effluent from SRB bioreactor (containing sulfide) which results in metal precipitation with less excessive sulfide for further treatment. Moreover, in metal sulfide precipitation process, process control can be used for pH and metal/sulfide stochiochemistry adjustment to provide the favorable conditions for selective metal precipitation (Sampaio et al., 2009; Veeken and Rulkens, 2003; Veeken et al., 2003).

- Selection of plant species

There are many plant species with hyper accumulation abilities and have economic values. *Helianthus annuus* (sunflower), *Nicotiana tabacum* (tobacco), *Vetiveria zizanioides* (vetiver grass), *Eichhornia crassipes* (water hyacinth) and *Nerium oleander* (oleander) are examples of hyper accumulators used in phytoremediation and also have economic values such as ornamental purposes or essential oil extraction (Chen et al., 2004; Danh et al., 2009; Jadia and Fulekar, 2008; Malik, 2007; Prasad and Freitas, 2003; Yoshihara et al., 2014). As a consequence, the plants harvested from the treatment sites can be sold or the extracted metal can be used.

7.5 General discussion

The management of e-waste in many countries is still very complicated and in the stage of infancy. Some countries have already applied certain regulations like extended producer responsibility (EPR) in Switzerland (Khetriwal et al., 2009) and Canada (McKerlie et al., 2006). Some developing countries have proposed some measures, but they still lack practical application due to lack of infrastructural support and an improperly functioning take back system (Kojima et al., 2009; Nnorom and Osibanjo, 2008b; Osibanjo and Nnorom, 2007). This is still an issue the stake holders need to take into consideration. Not only e-waste recycling will generate work opportunity for people, but it will also help to develop sustainable resource management and waste reduction. Bigum et al. (2012) has carried out life cycle analysis (LCA) and found that the recovery of metals from e-waste shows significant level in environmental and resource saving.

Although some of the commercially available e-waste treatment technologies may not be cost-effective, e-waste recycling and metal recovery should still be carried out with high priority. Considering their environmental impact assessment (EIA) and health risk assessment (HRA), making e-waste recycling legal and to have a proper control system will benefit the society in the long term. An example of long term benefit is reduced heavy metal exposure to humans. Lead (Pb) is one of the most toxic metals present in e-waste (Guo et al., 2010) which is considered to be toxic at any measurable concentration, especially to human's central nervous system (Needleman, 2004; Needleman and Bellinger, 2001).

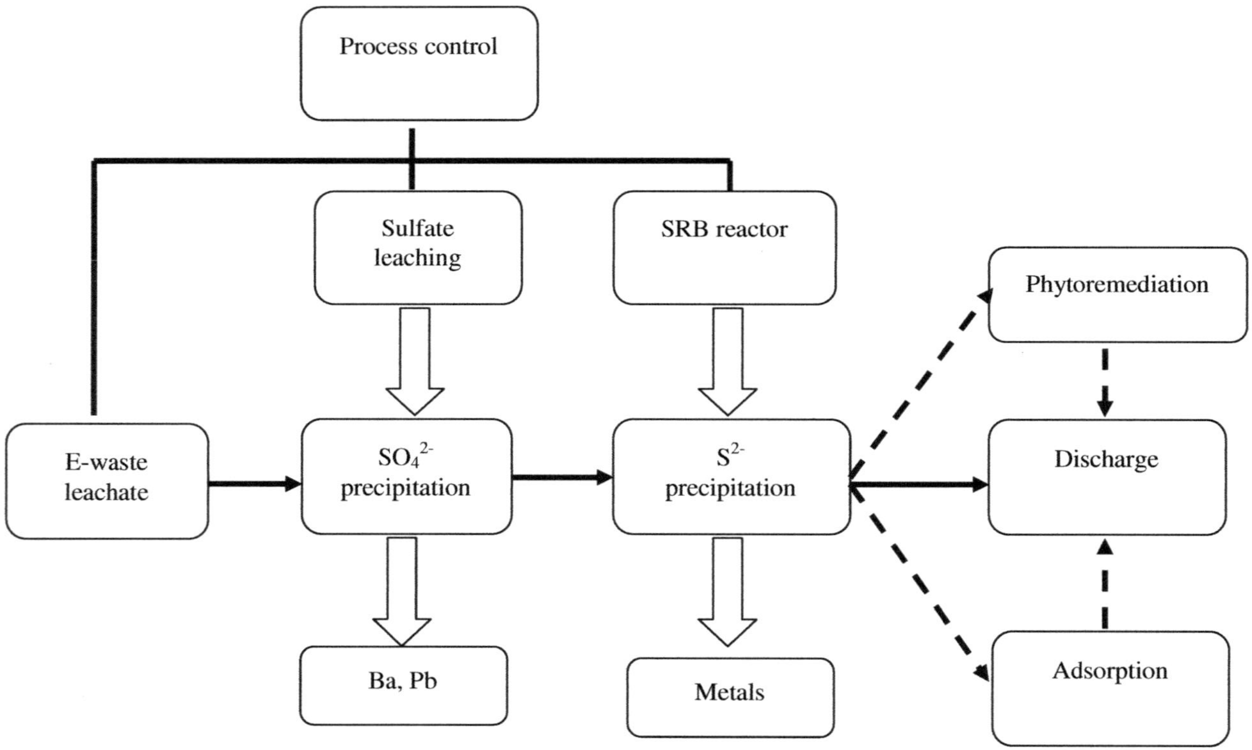

Figure 7.1 Combination system for metal removal and recovery

In addition to human, aquatic ecosystems will also get affected from the improper treatment of e-waste (Dagan et al., 2007), which will affect human in terms of food supply. The tragedy of Minamata in Japan is a typical example which was caused by Hg contaminated water reservoir (Kudo et al., 1998).

With rise in scientific and technological innovations, new technologies will certainly reduce the cost of the e-waste treatment processes. The utilization of the recovered metals is another issue which needs to be considered and emphasized. In order to promote this idea, the government should also take part in providing incentives such as tax reduction or subsidies for industries that uses recovered (secondary) metals as their raw materials.

References

Alazard, D., Dukan, S., Urios, A., Verhé, F., Bouabida, N., Morel, F., Thomas, P., Garcia, J.-L. and Ollivicr, B., 2003. *Desulfovibrio hydrothermalis sp. nov.*, a novel sulfate-reducing bacterium isolated from hydrothermal vents. Int J Syst Evol Microbiol, 53(1): 173-178.

Andrews, D., Raychaudhuri, A. and Frias, C., 2000. Environmentally sound technologies for recycling secondary lead. J. Power Sources, 88(1): 124-129.

Babu, B.R., Parande, A.K. and Basha, C.A., 2007. Electrical and electronic waste: a global environmental problem. Waste Manage. Res., 25(4): 307-318.

Barrett, E.L. and Clark, M.A., 1987. Tetrathionate reduction and production of hydrogen sulfide from thiosulfate. Microbiol. Rev., 51(2): 192-205.

Benatti, C.T., Tavares, C.R.G. and Lenzi, E., 2009. Sulfate removal from waste chemicals by precipitation. J. Environ. Manage., 90(1): 504-511.

Bigum, M., Brogaard, L. and Christensen, T.H., 2012. Metal recovery from high-grade WEEE: A life cycle assessment. J. Hazard. Mater., 207-208(0): 8-14.

Bijmans, M.F.M., Buisman, C.J.N., Meulepas, R.J.W. and Lens, P.N.L., 2011. Sulfate reduction for inorganic waste and process water treatment. Compr. Biotechnol.: 435-446.

Castor, S.B. and Hedrick, J.B., 2006. Rare earth elements. Industrial Minerals volume, 7th edition: Society for Mining, Metallurgy, and Exploration, Littleton, Colorado: 769-792.

Chen, Y., Shen, Z. and Li, X., 2004. The use of vetiver grass (*Vetiveria zizanioides*) in the phytoremediation of soils contaminated with heavy metals. Appl. Geochem., 19(10): 1553-1565.

Choi, J.-Y., Kim, D.-S. and Lim, J.-Y., 2006. Fundamental features of copper ion precipitation using sulfide as a precipitant in a wastewater system. J. Environ. Sci. Health, Part A, 41(6): 1155-1172.

Cropper, M. and Griffiths, C., 1994. The interaction of population growth and environmental quality. Am. Econ. Rev.: 250-254.

Cui, J. and Zhang, L., 2008. Metallurgical recovery of metals from electronic waste: A review. J. Hazard. Mater., 158(2-3): 228-256.

Dagan, R., Dubey, B., Bitton, G. and Townsend, T., 2007. Aquatic toxicity of leachates generated from electronic devices. Arch Environ Contam Toxicol, 53(2): 168-173.

Danh, L.T., Truong, P., Mammucari, R., Tran, T. and Foster, N., 2009. Vetiver grass, *Vetiveria zizanioides*: a choice plant for phytoremediation of heavy metals and organic wastes. Int. J. Phytoremediation, 11(8): 664-691.

Deng, X., Huang, J., Rozelle, S. and Uchida, E., 2008. Growth, population and industrialization, and urban land expansion of China. J. Urban Econ., 63(1): 96-115.

Dodson, J.R., Hunt, A.J., Parker, H.L., Yang, Y. and Clark, J.H., 2012. Elemental sustainability: Towards the total recovery of scarce metals. Chem. Eng. Process., 51(0): 69-78.

Downs, A.J., 1993. *Chemistry of aluminium, gallium, indium, and thallium*. Springer. New York.

Du, X. and Graedel, T., 2011. Uncovering the global life cycles of the rare earth elements. Scientific reports, 1.

Dubey, S.P. and Gopal, K., 2007. Adsorption of chromium (VI) on low cost adsorbents derived from agricultural waste material: a comparative study. J. Hazard. Mater., 145(3): 465-470.

Dunn, I.J., Heinzle, E., Ingham, J. and Přenosil, J.E., 2005. Automatic Bioprocess Control Fundamentals, Biological Reaction Engineering. Wiley, New York. Wiley, New York.

Ehrlich, P.R. and Holdren, J.P., 1971. Impact of population growth. Science, 171 (3977): 1212-1217.

Fu, J., Zhou, Q., Liu, J., Liu, W., Wang, T., Zhang, Q. and Jiang, G., 2008. High levels of heavy metals in rice (*Oryza sativa* L.) from a typical e-waste recycling area in southeast China and its potential risk to human health. Chemosphere, 71(7): 1269-1275.

Greber, J.F., 2000. Gallium and Gallium Compounds, Ullmann's Encyclopedia of Industrial Chemistry. Wiley, New York.

Guo, Y., Huo, X., Li, Y., Wu, K., Liu, J., Huang, J., Zheng, G., Xiao, Q., Yang, H., Wang, Y., Chen, A. and Xu, X., 2010. Monitoring of lead, cadmium, chromium and nickel in placenta from an e-waste recycling town in China. Sci. Total Environ., 408(16): 3113-3117.

Ha, N.N., Agusa, T., Ramu, K., Tu, N.P.C., Murata, S., Bulbule, K.A., Parthasaraty, P., Takahashi, S., Subramanian, A. and Tanabe, S., 2009. Contamination by trace elements at e-waste recycling sites in Bangalore, India. Chemosphere, 76(1): 9-15.

Havlik, T., Orac, D., Petranikova, M., Miskufova, A., Kukurugya, F. and Takacova, Z., 2010. Leaching of copper and tin from used printed circuit boards after thermal treatment. J. Hazard. Mater., 183(1-3): 866-873.

Hlabela, P., Maree, J. and Bruinsma, D., 2007. Barium carbonate process for sulphate and metal removal from mine water. Mine Water Environ, 26(1): 14-22.

Holdren, J.P. and Ehrlich, P.R., 1974. Human population and the global environment: Population growth, rising per capita material consumption, and disruptive technologies have made civilization a global ecological force. Am. Sci.: 282-292.

Jadia, C.D. and Fulekar, M.H., 2008. Phytoremediation: the application of vermicompost to remove zinc, cadmium, copper, nickel and lead by sunflower plant. Environ. Eng. Manage. J., 7(5): 547-558.

Jang, Y.-C. and Townsend, T., 2001. Sulfate leaching from recovered construction and demolition debris fines. Adv. Environ. Res., 5(3): 203-217.

Jørgensen, B.B. and Bak, F., 1991. Pathways and microbiology of thiosulfate transformations and sulfate reduction in a marine sediment (Kattegat, Denmark). Appl. Environ. Microb., 57(3): 847-856.

Kadirvelu, K., Thamaraiselvi, K. and Namasivayam, C., 2001. Removal of heavy metals from industrial wastewaters by adsorption onto activated carbon prepared from an agricultural solid waste. Bioresour. Technol., 76(1): 63-65.

Kaksonen, A.H. and Puhakka, J.A., 2007. Sulfate reduction based bioprocesses for the treatment of acid mine drainage and the recovery of metals. Eng. Life Sci., 7(6): 541-564.

Kamsonlian, S., Suresh, S., Majumder, C. and Chand, S., 2011. Characterization of banana and orange peels: biosorption mechanism. Int. J. Sci. Technol. Manage, 2(4): 1-7.

Khetriwal, D.S., Kraeuchi, P. and Widmer, R., 2009. Producer responsibility for e-waste management: key issues for consideration-learning from the Swiss experience. J. Environ. Manage., 90(1): 153-165.

Kim, E.-y., Kim, M.-s., Lee, J.-c., Jeong, J. and Pandey, B.D., 2011. Leaching kinetics of copper from waste printed circuit boards by electro-generated chlorine in HCl solution. Hydrometallurgy, 107(3-4): 124-132.

Kojima, M., Yoshida, A. and Sasaki, S., 2009. Difficulties in applying extended producer responsibility policies in developing countries: case studies in e-waste recycling in China and Thailand. J Mater Cycles Waste Manag, 11(3): 263-269.

Kudo, A., Fujikawa, Y., Miyahara, S., Zheng, J., Takigami, H., Sugahara, M. and Muramatsu, T., 1998. Lessons from Minamata mercury pollution, Japan-after a continuous 22 years of observation. Water Sci. and Technol., 38(7): 187-193.

Lee, J.-C., Koyanaka, S., Lee, M., Ohya, H. and Endoh, S., 1997. Recovery of copper, tin and lead from the spent printed circuit boards (PCBs) by the shape separation method. Shigen-to-Sozai, 113(5): 357-362.

Lens, P. and Kuenen, J., 2001. The biological sulfur cycle: novel opportunities for environmental biotechnology. Water Sci.Technol., 44(8): 57-66.

Lens, P.N.L. and Pol, L.H., 2000. *Environmental Technologies to Treat Sulfur Pollution: Principles and Engineering*. IWA Publishing, London, United Kingdom.

Lewis, A.E., 2010. Review of metal sulphide precipitation. Hydrometallurgy, 104(2): 222-234.

Luo, C., Liu, C., Wang, Y., Liu, X., Li, F., Zhang, G. and Li, X., 2011. Heavy metal contamination in soils and vegetables near an e-waste processing site, south China. J. Hazard. Mater., 186(1): 481-490.

Malik, A., 2007. Environmental challenge *vis a vis* opportunity: The case of water hyacinth. Environ. Int., 33(1): 122-138.

Maree, J., Hlabela, P., Nengovhela, R., Geldenhuys, A., Mbhele, N., Nevhulaudzi, T. and Waanders, F., 2004. Treatment of mine water for sulphate and metal removal using barium sulphide. Mine Water Environ, 23(4): 195-203.

McKerlie, K., Knight, N. and Thorpe, B., 2006. Advancing extended producer responsibility in Canada. J. Clean. Prod., 14(6-7): 616-628.

Mecucci, A. and Scott, K., 2002. Leaching and electrochemical recovery of copper, lead and tin from scrap printed circuit boards. J. Chem. Techno. Biotechnol., 77(4): 449-457.

Merian, E., Anke, M., Ihnat, M. and Stoeppler, M., 2004. Elements and their compounds in the environment: occurrence, analysis and biological relevance. Wiley, New york.

Mersmann, A., 1999. Crystallization and precipitation. Chem. Eng.Process., 38(4-6): 345-353.

Mohan, D. and Singh, K.P., 2002. Single-and multi-component adsorption of cadmium and zinc using activated carbon derived from bagasse-an agricultural waste. Water Res., 36(9): 2304-2318.

Moskalyk, R.R., 2003. Gallium: the backbone of the electronics industry. Miner. Eng., 16(10): 921-929.

Needleman, H., 2004. Lead poisoning. Annu. Rev. Med., 55: 209-222.

Needleman, H.L. and Bellinger, D., 2001. Studies of lead exposure and the developing central nervous system: a reply to Kaufman. Arch. Clin. Neuropsych., 16(4): 359-374.

Nnorom, I.C. and Osibanjo, O., 2008a. Electronic waste (e-waste): Material flows and management practices in Nigeria. Waste Manage., 28(8): 1472-1479.

Nnorom, I.C. and Osibanjo, O., 2008b. Overview of electronic waste (e-waste) management practices and legislations, and their poor applications in the developing countries. Resour. Conser. Recy., 52(6): 843-858.

Orhan, Y. and Büyükgüngör, H., 1993. The removal of heavy metals by using agricultural wastes. Water Sci. Technol., 28(2): 247-255.

Osibanjo, O. and Nnorom, I.C., 2007. The challenge of electronic waste (e-waste) management in developing countries. Waste Manage. Res., 25(6): 489-501.

Peters, R.W., Chang, T.-K. and Ku, Y., 1984. Heavy metal crystallization kinetics in an MSMPR crystallizer employing sulfide precipitation. AIChE Symp., 240 (80): 55-75.

Pikaar, I., Meulepas, R.J., Jagersma, C.G., Stams, A.J. and Lens, P.N., 2013. Thiosulfate removal by disproportionation in anaerobic thermophilic bioreactors, G16 Conference, pp. 83.

Pind, P., Angelidaki, I., Ahring, B., Stamatelatou, K. and Lyberatos, G., 2003. Monitoring and Control of Anaerobic Reactors. In: B. Ahring et al. (Eds.), *Biomethanation II. Advances in Biochemical Engineering/Biotechnology*. Springer Berlin Heidelberg, pp. 135-182.

Prasad M.N.V. and Freitas H.M.O, 2003. Metal hyperaccumulation in plants: biodiversity prospecting for phytoremediation technology. Electro. J. Biotechnol. 6: 285-321

Reck, B.K. and Graedel, T., 2012. Challenges in metal recycling. Science, 337(6095): 690-695.

Roberts, M.C., 1990. Predicting metal consumption: The case of US steel. Resour. Policy, 16(1): 56-73.

Robinson, B.H., 2009. E-waste: An assessment of global production and environmental impacts. Sci. Total Environ., 408(2): 183-191.

Sampaio, R.M.M., Timmers, R.A., Xu, Y., Keesman, K.J. and Lens, P.N.L., 2009. Selective precipitation of Cu from Zn in a pS controlled continuously stirred tank reactor. J. Hazard. Mater., 165(1-3): 256-265.

Schwarzer, S., De Bono, A., Giuliani, G., Kluser, S. and Peduzzi, P., 2005. E-waste, the hidden side of IT equipment's manufacturing and use. http://www.grid.unep.ch/product/publication/download/ew_ewaste.pdf (accessed on 29 September 2014)

Sepúlveda, A., Schluep, M., Renaud, F.G., Streicher, M., Kuehr, R., Hagelüken, C. and Gerecke, A.C., 2010. A review of the environmental fate and effects of hazardous substances released from electrical and electronic equipments during recycling: Examples from China and India. Environ. Impact Assess. Rev., 30(1): 28-41.

Shen, C., Chen, Y., Huang, S., Wang, Z., Yu, C., Qiao, M., Xu, Y., Setty, K., Zhang, J., Zhu, Y. and Lin, Q., 2009. Dioxin-like compounds in agricultural soils near e-waste recycling sites from Taizhou area, China: Chemical and bioanalytical characterization. Environ. Int., 35(1): 50-55.

Špaldon, T., Hančuľák, J., Šestinová, O., Findoráková, L. and Kurbel, T., 2013. Utilization of various barium compounds as sulphates remover, Waste: 164-169.

Sud, D., Mahajan, G. and Kaur, M., 2008. Agricultural waste material as potential adsorbent for sequestering heavy metal ions from aqueous solutions - a review. Bioresour. Technol., 99(14): 6017-6027.

Tabak, H.H., Scharp, R., Burckle, J., Kawahara, F.K. and Govind, R., 2003. Advances in biotreatment of acid mine drainage and biorecovery of metals: 1. Metal precipitation for recovery and recycle. Biodegradation, 14(6): 423-436.

Tuncuk, A., Stazi, V., Akcil, A., Yazici, E.Y. and Deveci, H., 2012. Aqueous metal recovery techniques from e-scrap: Hydrometallurgy in recycling. Miner. Eng., 25(1): 28-37.

Veeken, A. and Rulkens, W., 2003. Innovative developments in the selective removal and reuse of heavy metals from wastewaters. Water Sci. Technol., 47(10): 9-16.

Veeken, A.H.M., Akoto, L., Hulshoff Pol, L.W. and Weijma, J., 2003. Control of the sulfide (S^{2-}) concentration for optimal zinc removal by sulfide precipitation in a continuously stirred tank reactor. Water Res., 37(15): 3709-3717.

Wang, D., Cai, Z., Jiang, G., Leung, A., Wong, M.H. and Wong, W.K., 2005. Determination of polybrominated diphenyl ethers in soil and sediment from an electronic waste recycling facility. Chemosphere, 60(6): 810-816.

WHO, 2010. Childhood lead poisoning, World Health Organization. Geneva: WHO.

Widmer, R., Oswald-Krapf, H., Sinha-Khetriwal, D., Schnellmann, M. and Böni, H., 2005. Global perspectives on e-waste. Environ. Impact Assess. Rev., 25(5): 436-458.

Wong, M.H., Wu, S.C., Deng, W.J., Yu, X.Z., Luo, Q., Leung, A.O.W., Wong, C.S.C., Luksemburg, W.J. and Wong, A.S., 2007. Export of toxic chemicals - A review of the case of uncontrolled electronic-waste recycling. Environ. Pollut., 149(2): 131-140.

Wu, C.M.L. and Wong, Y.W., 2007. Rare-earth additions to lead-free electronic solders. J. Mater. Sci, 18(1-3): 77-91.

Yang, H., Liu, J. and Yang, J., 2011. Leaching copper from shredded particles of waste printed circuit boards. J. Hazard. Mater., 187(1-3): 393-400.

Yellishetty, M., Mudd, G.M. and Ranjith, P.G., 2011. The steel industry, abiotic resource depletion and life cycle assessment: a real or perceived issue?. J. Clean. Prod., 19(1): 78-90.

Yoshihara, T., Suzui, N., Ishii, S., Kitazaki, M., Yamazaki, H., Kitazaki, K., Kawachi, N., Yin, Y.-G., Ito-Tanabata, S., Hashida, S.-N., Shoji, K., Shimada, H., Goto, F. and Fujimaki, S.H.U., 2014. A kinetic analysis of cadmium accumulation in a Cd hyper-accumulator fern, *Athyrium yokoscense* and tobacco plants. Plant Cell Environ., 37(5): 1086-1096.

Zhang, Q., Zhou, T., Xu, X., Guo, Y., Zhao, Z., Zhu, M., Li, W., Yi, D. and Huo, X., 2011. Downregulation of placental S100P is associated with cadmium exposure in Guiyu, an e-waste recycling town in China. Sci. Total Environ., 410-411(0): 53-58.

Zheng, L., Wu, K., Li, Y., Qi, Z., Han, D., Zhang, B., Gu, C., Chen, G., Liu, J. and Chen, S., 2008. Blood lead and cadmium levels and relevant factors among children from an e-waste recycling town in China. Environ. Res., 108(1): 15-20.

Netherlands Research School for the
Socio-Economic and Natural Sciences of the Environment

D I P L O M A

For specialised PhD training

The Netherlands Research School for the
Socio-Economic and Natural Sciences of the Environment
(SENSE) declares that

Suthee Janyasuthiwong

born on 1 October 1986 in Bangkok, Thailand

has successfully fulfilled all requirements of the
Educational Programme of SENSE.

Delft, 3 July 2015

the Chairman of the SENSE board

Prof. dr. Huub Rijnaarts

the SENSE Director of Education

Dr. Ad van Dommelen

The SENSE Research School has been accredited by the Royal Netherlands Academy of Arts and Sciences (KNAW)

KONINKLIJKE NEDERLANDSE
AKADEMIE VAN WETENSCHAPPEN

The SENSE Research School declares that Mr Suthee Janyasuthiwong has successfully fulfilled all requirements of the Educational PhD Programme of SENSE with a work load of 51.2 EC, including the following activities:

SENSE PhD Courses

- Environmental Research in Context (2012)
- Speciation and Bioavailability (2013)
- Research in Context Activity: Organising summer school on 'Contaminated sediments, characterisation and remediation', Delft (2013)

Other PhD and Advanced MSc Courses

- Nanotechnology for water and wastewater, UNESCO-IHE, Delft (2011)
- Wastewater treatment technologies and modelling, University Paris Est (2012)
- Summer School: Biofilms for technology, UNESCO-IHE, Delft (2012)
- Project and time management, Wageningen University (2013)
- ETeCoS[3] summer school 'Contaminated sediments, characterisation and remediation', UNESCO-IHE, Delft (2013)
- Cleaner production & the water cycle, UNESCO-IHE, Delft (2014)

External training at a foreign research institute

- Contaminated soils. From characterization to remediation, EteCoS[3] Summer school, University Paris Est, France (2012)
- EteCoS[3] summer school 'Biological treatment of solid waste', Cassino and Gaeta, Italy (2014)

Management and Didactic Skills Training

- Supervising two MSc students, UNESCO-IHE, Delft, The Netherlands (2012-2014)

Selection of Oral Presentations

- *Biogenic Sulfide Production and Selective Metal Precipitation at Low pH for Semiconductor Wastewater Treatment*. SENSE Science Market: Toward a Biobased Economy, 25 October 2012, Den Haag, The Netherlands
- *Hydrobiometallurgical processes for Metal Recovery from Desktop Computer Printed Circuit Boards*. UNESCO-IHE PhD week: Urban sustainability, 29 September 2014, Delft, The Netherlands

SENSE Coordinator PhD Education

Dr. ing. Monique Gulickx

Curriculum Vitae

Suthee Janyasuthiwong was born on 1st October 1986 in Bangkok (Thailand). He obtained his diploma in Bachelor of Science in Biology (Botany) in 2008 from Kasetsart University (Bangkok, Thailand) with second class honors under the fellowship from Science Achievement Scholarship of Thailand (SAST) by Human Resource Development in Science Project. From 2008 to 2010, he pursued his post graduate study with specialization in Environmental Technology Toxicology and Management, a collaboration between the Chulabhorn Research Institute (CRI), Mahidol University (MU) and Asian Institute of Technology (AIT) under the fund from S&T Postgraduate Education and Research Development Office (PERDO) and Royal Thai Government fellowship. From November 2011, he started his doctoral degree with the joint doctorate program supported by Erasmus Mundus Joint Doctorate in Environmental Technologies for Contaminated Solids, Solis, and Sediments (ETeCoS^3). His research was carried out majorly at UNESCO-IHE Institute for Water Education (the Netherlands) with the academic mobility to University of Cassino and Southern Lazio (Italy). The main focus of his research was on metal removal and recovery from the waste streams.

Publications and Presentations

Publications

- **Janyasuthiwong S.**, Rene E.R., Esposito G. and Lens P.N.L. (2015). Effect of pH on the performance of sulfate and thiosulfate fed sulfate reducing inverse fluidized bed reactors. ***J. Environ. Eng.,*** *In press*. DOI: 10.1061/(ASCE)EE.1943-7870.0001004
- **Janyasuthiwong S.**, Phiri S.M., Kijjanapanich P., Rene E.R., Esposito G. and Lens P.N.L. (2015). Copper, lead and zinc removal from metal contaminated wastewater by adsorption onto agricultural wastes. ***Environ Technol.,*** *In Press*. DOI: 10.1080/09593330.2015.1053537
- **Janyasuthiwong S.**, Rene E.R., Esposito G. and Lens P.N.L. (2015). Effect of pH on Cu, Ni and Zn removal by biogenic sulfide precipitation in an inversed fluidized bed bioreactor. ***Hydrometallurgy***, *In Press*
- **Janyasuthiwong S.**, Ugas R., Rene E.R., Alessandra C., Esposito G. and Lens P.N.L. (2015). Effect of operational parameters on the leaching efficiency and recovery of heavy metals from computer printed circuit boards. ***J. Chem. Technol. Biotechnol,*** *Accepted* DOI: 10.1002/jctb.4798
- **Janyasuthiwong S.**, Rene E.R., Esposito G. and Lens P.N.L. (2015). Metal removal and resource recovery technologies from waste stream. In: *Sustainable Technologies for Heavy Metal Removal from Soils, Solid Wastes and Wastewater*, Lens, P. N. L., Rene, E. R., Lewis, A., Sahinkaya, E. (Eds.) Book series: **Environmental Chemistry for a Sustainable World** (Eds: Lichtfouse, E., Schwarzbauer, J., Robert, D), Springer, *Under revision*

Presentations

- Biogenic sulfide production and selective metal precipitation at low pH for semiconductor wastewater treatment. Microbial Sulfur Metabolism, E.M.B.O. workshop. 15 - 18 April 2012. Noordwijkerhout, the Netherlands. (Poster)
- Biogenic sulfide production and selective metal precipitation at low pH for semiconductor wastewater treatment. UNESCO-IHE PhD week: Managing Water Resources in a Changing World, Delft, the Netherlands.

- Biogenic sulfide production and selective metal precipitation at low pH for semiconductor wastewater treatment. SENSE Science Market: Toward a Biobased Economy, 25 October 2012, Den Haag, the Netherlands
- Biogenic sulfide production and metal precipitation in an inversed fluidized bed bioreactor. 3rd International Conference on Research Frontiers in Chalcogen Cycle Science & Technology. 27- 28 May 2013. Delft, the Netherlands. (Poster)
- Biogenic sulfide precipitation for metal removal and recovery for electronic waste leachate treatment. 4th Granada International School of Crystallization. 26 - 30 May 2014. Granada, Spain. (Poster)
- Hydrobiometallurgical processes for metal recovery from desktop computer printed circuit boards. UNESCO-IHE PhD week: Urban sustainability 2014, Delft, the Netherlands.

For Product Safety Concerns and Information please contact our EU representative GPSR@taylorandfrancis.com Taylor & Francis Verlag GmbH, Kaufingerstraße 24, 80331 München, Germany

Batch number: 10726324

Printed by Printforce, the Netherlands